To

Dr Murrison,

The Story of Electricity

From

Tom

28/6/11

The Story of Electricity

by

T. S. M. MacLean

DB

DIADEM BOOKS

The Story of Electricity
All Rights Reserved. Copyright © 2011 T.S.M. MacLean

Published by Diadem Books
Distribution coordination by Spiderwize

For information, please contact:

Diadem Books
16 Lethen View
Tullibody
Alloa
FK10 2GE
Scotland UK

www.diadembooks.com

ISBN: 978-1-908026-04-0

Personal communion
in memory of J.M.

Table of Contents

Preface

THIS IS A BOOK which unusually includes looking at the religious beliefs of people who were not theologians, but who have demonstrated clear thinking ability in their own field of study. All the people to be considered are scientists, but to avoid controversy over evolution, none are biologists.

Attention is restricted to one branch of physics, viz, magnetism and electricity, covering the time scale between c.350 B.C. and c.1950 A.D. Within this branch, since every change in a magnetic field automatically produces an electric field, the thrust of the book, including its title, is directed mainly towards electricity.

Within the specified time scale it will be clear that thousands of people will have contributed to our present day knowledge of this subject. Only a small number can be included, and reference is made to just a few score of these. But the information provided is necessarily unequally distributed. Some have had to be treated rather summarily because of the absence of sufficient biographical information about them, even in their own language. Others have had rather reticent biographers who say little or nothing about the religious beliefs of their subjects, and others appear to have kept such knowledge to themselves, believing that such matters are essentially private.

Accepting such limitations and proceeding chronologically, Chapter One includes the story of Hildegaard of Bingen, abbess, poetess, musician and theologian in the 12th Century, who has furnished us with the saying, "The gifts of God do not grow old like old clothes, but may grow into things of beauty." She used a magnet, moistened with the patient's saliva and drawn across his forehead, to attempt a cure for insanity.

In the 13th Century, Albert the Great, eventually Rector at Cologne, asserted there were two certainties in life. The first is the certainty of revealed religion. The second is the certainty of one's own faith in the truth of one's personal experience. Albert was the only mediaeval philosopher to make commentaries on all the works of Aristotle, and enriched science by his own scientific observations. He was a Dominican Friar, who therefore practised zeal for souls as his aim in life, to be distinguished from a Grey Friar, whose practice in life was personal poverty.

Another claim to fame of Albert the Great was his ability to attract students to Cologne. Among these was Thomas Aquinas. He was known originally to his fellow students as "The Dumb Ox", because he was big physically and did not speak when speaking was not required. But he eventually acquired such a reverence in the Roman Catholic Church that his *Summa Theologica* writings were placed on the altar at the Council of Trent (1545-1563) along with the Sacred Scriptures and the Papal Decrees. One of Thomas's sayings was, "The beatitude of spiritual creatures is the central purpose of creation. Thus the physical universe is anthropomorphic in a way that remained unknown to the Arabs and Greeks."

Within one generation following Thomas Aquinas, i.e. in the 14th Century, there came Duns Scotus, from Duns in Berwickshire. He saw no difficulty in accepting the theory of action at a distance, citing the analogy of the moon exerting an influence on the tides, like the influence he said of lodestone on iron. This appears to have had no effect for hundreds of years on the belief in the existence of an aether surrounding the earth, associated with continuous forces on neighbouring particles. His tomb in Cologne bears the inscription in Latin: "Scotia brought me forth, England sustained me, France taught me, Cologne holds me."

Scotus, as a Franciscan Friar, took the view that the primary concern of Theology is God, whereas Philosophy treats God only as the first cause of things. Regarding Theology as a science, Scotus saw it from the practical point of view as being concerned with saving souls through Revelation. He argued that a person may know by faith, with absolute certainty, that the human soul both exists and is immortal. In contrast to this, reason can only postulate the existence of such a soul. One of the other doctrines which he defended was that of the Immaculate Conception of Mary, i.e. the belief that she was conceived without Original Sin. This view was defined by Pope Pius IX in 1854, as a dogma of the Roman Catholic Church.

The discovery of America by Christopher Columbus in the 15th Century was the lodestone's greatest triumph. It is believed that Columbus had a vision of reaching Japan and China by sailing West. In this belief he would have found support from 2 Esdras 6 v 42 where we read that on Day Three of Creation "You ordered the waters to collect on a seventh part of the earth; the other six parts you made into dry land, and from it kept some to be sown and tilled for your service. Your Word went forth and at once the work was done." This quotation is from the Apocrypha, not the Old Testament, and is in error. But it is easy to see how a religious person, like Columbus, believed this error.

In the 16th Century William Gilbert was a Court Physician to Queen Elizabeth of England. He is remembered today chiefly for his scientific rather than his medical work. His emphasis on experimental measurements stands out as the dominant factor in his book *De Magnete*, which was a turning point in the history of scientific literature. No longer was it sufficient to say what Aristotle had said, because Gilbert had carried out careful measurements himself. His work included both magnetic and electric studies. But in covering such a vast field, it is not surprising that when he attempted to generalise his experimental results he should have fallen into error. Thus he says, "The earth seeks the sun, turning from him and following him by her magnetic energy." He believed that every magnetic body, like the earth, imitated a soul, providing the power for its "rotational movement". This view was unacceptable to those who did not accept the analogy of the earth possessing a soul, and among these was Renee Descartes.

On the 10thNovember 1619 Renee Descartes, the son of a French magistrate, then aged twenty-three, had a religious vision which clarified for him what his objective in life should be. His life should be devoted to physics based on geometry. To give thanks for this vision he sold his estate and went on a pilgrimage. After living in Holland for some years he wrote his first major work, *Le Monde* in 1633, which dealt with celestial and terrestrial physics, from a Copernican viewpoint. This was the year in which Galileo Galilei, on his knees, had to make his well-known confession, renouncing his Copernican views. Consequently Renee Descartes withheld his manuscript from publication, and it was not published until after his death. One of Descarte's religious sayings was, "The body is a mechanism of many parts. In contrast the soul and God are simple and indivisible." With regard to science he believed that all occurrences in nature could be predicted by calculation. But this view was not shared by others, who knew that the truths of physics have to be built up experimentally. Among these was the Irishman

Robert Boyle, also a 17th Century scientist, who, although primarily a chemist, contributed substantially in both magnetism and electricity.

Among Boyle's many religious sayings, the following three have been selected as typical:

(1) "The possibility of changing base metals into gold is analogous to the redemption of mankind in theology."
(2) "Religious doubts are like toothache, not fatal but very troublesome."
(3) "He alone loves God as much as he ought, that loving him as much as he can, strives to repair the deplored imperfection of that love, with an extreme regret at finding his love no greater."

The eighteenth century saw a growing interest in electricity. Along with it there was an accompanying increase in the number of religious sayings, from workers in that field. Thus, in America, Benjamin Franklin who showed that lightning was an electrical phenomenon, provided a number of couplets:

"Fear not death, for the sooner we die,
The longer shall we be immortal."

"Think of three things—Whence you came
Where you are going and to Whom you must account."

From Switzerland, Leonhard Euler, the outstanding mathematician of that century, has given a multitude of religious sayings, of which in this Preface only one will be included. It is:

"When a man addresses to God a prayer worthy of being heard, it must not be imagined that such a prayer came not to the knowledge of God till the moment it was formed. That prayer was clearly heard from all eternity. And if the Father of Mercies, deemed it worthy of being answered, He arranged the world expressly in favour of that prayer, so that the accomplishment should be a consequence of the natural course of events. It is thus that God answers the prayers of men, without working a miracle. Thus all our prayers have been already presented at the throne of the Almighty, and have been admitted into the plan of the Universe, in subservience to the infinite wisdom of the Creator."

Similarly in Germany, Georg Lichtenberg, the proposer of the A4 size of paper we almost all use today, has left us again with many sayings, of which the following four are included:

(1) "Hour glasses remind us not only of the passage of time, but also of the dust to which we shall come one day."
(2) "Any experiment through which new knowledge comes in a very concentrated form, is the work of God suddenly revealed to man's intellect."
(3) "The soul places the countenance around itself like a magnet does iron filings."
(4) "It has been observed long since that when the Spirit rises up, the body falls on its knees."

In the 19th Century Andre Marie Ampere came to be described as the Father of Electromagnetism. As a boy he was endowed with a phenomenal memory, with which he memorised "The Imitation of Christ". On his deathbed when a friend started to read from it he said, "Thank you, but I know it all by heart." It is recorded that his first Communion was one of the three most important events in his life. The other two were reading a biography of Descartes, which led him to Science, and the fall of the Bastille, which fixed his political allegiance.

Ampere formed a Society of Christians, in which he was given the task of showing that Christianity was Divinely inspired. He carried this out as an intellectual exercise, but he struggled throughout his life to achieve equilibrium between his heart and his intellect. One of his sayings is: "Doubt is the greatest torment that man has on earth." But he was a good Christian, sharing his accommodation in Paris with A.J. Fresnel who was an ill scientist of Jansenist persuasion, though Ampere himself always remained a Roman Catholic.

Another good Christian in the same 19th Century was Michael Faraday. Throughout his working life he was strengthened by the theological beliefs of the Sandemanian Church in which he had been brought up. They affected his response to the outside world. It provided him with the assurance that there was a distinct plan behind events. If he fell ill his fellow church members rallied to his support. And at the time when his church called him to be an elder, his response indicated that although he accepted this as part of God's plan for him, he trembled at the thought of his future Day of Judgment, lest he then be found to be personally unworthy.

In parallel with Michael Faraday is the life of Joseph Henry, the inventor of the telegraph system in America. Asked later why he did not patent this invention, he replied, "I did not then consider it compatible with the dignity of science to confine benefits which might be derived from it to the exclusive use of any individual." Apart from his electromagnetic discoveries which overlapped those of Faraday, perhaps his greatest claim to fame was being Secretary of the famous Smithsonian Institute, which he directed for the second half of his life.

In Germany, Hermann Helmholtz (1841-1894) as a young researcher worked in a small group headed by an eminent physiologist. His comment on this experience was, "Whoever comes into contact with men of the first rank has an altered scale of values in life." Later he himself wrote a paper "On the Physiological Causes of Harmony in Music", which still stands as an example of pedagogic clarity. He was baptised into the Lutheran Church, but he never courted extremes in religious matters. Rather by education and conviction, he was religious in the noblest sense. His outstanding contribution to electricity was made in a lecture he gave to the Fellows of the Chemical Society in 1881. Lord Kelvin has described this lecture as "an epoch-making monument of the progress of Natural Philosophy in the nineteenth century, in virtue of the declaration, then first made, that electricity consists of atoms. Before that time atomic theories of electricity had been noticed and rejected by Faraday and Maxwell, and probably by many other philosophers and workers; but certainly accepted by none".

As a small boy William Thomson (later to become Lord Kelvin) attended a small school kept by the Minister of the Secession Presbyterian Church in Northern Ireland, to learn classics and mathematics. At age sixteen he took Fourier's *Analytic Theory of Heat* out of the Glasgow University Library, and in a fortnight had mastered it. One year later he transferred to Cambridge University, where he became Second Wrangler and Smith's Prizeman. His marriage to Margaret Crum in 1852 was to a girl of a deeply religious nature, an original poet and also a translator from German poets. After the Scottish Disruption of 1843 they associated themselves with the Free Church of Scotland.

James Clerk Maxwell was gifted with an excellent memory. He was able to recite the whole of the 119th Psalm, and it is said that he was able to identify from which Psalm almost any quotation came. At Edinburgh Academy his fellow pupils initially called him "Dafty", which is reminiscent of the sobriquet given to Thomas Aquinas, who in Cologne was called "The Dumb Ox". Later in Cambridge at age twenty-two he wrote "A Student's Evening Hymn", which shows maturity beyond his years, in parallel with his scientific progress. One of the religious influences on his family came from Charles Mackenzie, six years older than James. He also attended both Edinburgh Academy, and Cambridge University where he became Second Wrangler also. Later in life he was consecrated Bishop, served under David Livingstone in Malawi, before dying from Malaria at the age of thirty-seven.

Max Planck (1858-1947) took up the role of preaching after his scientific career had ended. From 1920 until his death, he had been an elder in his Lutheran church in Berlin, and he was accustomed to saying grace at meal times. His rallying cry for both religion and science was "On to God".

Planck defines religion as the link which binds man to God. It is based on our humility before Him. But there is also an active desire on our part to be in harmony with Him, and to be protected from visible and invisible dangers which surround. The result of this is to produce an inner peace of mind and soul, and a trusting faith in His omnipotence and benevolence. But then Planck points out that the significance of religion goes beyond the individual. It applies also to the larger community and even to the whole world. In religion God is the starting point, whereas in science He is the crown of the structure. Science wants man to learn, but religion wants him to act. For this religious action there must be a direct link to God. With this link there comes an inner firmness and peace of mind, which is the highest boon of life.

Following Planck, Julian Schwinger (1918-1994) was a Jew, as was Richard Feynman, both of whom were born and educated in New York. Both also became Nobel Laureates, and Schwinger himself supervised three students who later themselves also became Nobel Laureates. But it was Feynman who became a world known teacher of electricity. This may have been associated with the loss, through tuberculosis, of his first wife early in his career, and his inability to accept the Catholic faith of a second girl whom he later wished to marry. Feynman died in 1988.

The approaches of Schwinger and Feynman to the problem of how electrically charged particles and electromagnetic fields interact were fundamentally different. Schwinger used a field theory approach, following earlier work of Dirac in Cambridge, but Feynman stuck to a particle approach. Although the results appeared initially to be different, it was shown later by Dyson that they agreed completely with each other.

CHAPTER ONE

THE EARLY HISTORY OF ELECTRICITY

(c.350 B.C. – c.1550 A.D.)

T HE EFFECTS OF ELECTRICITY were known in the pre-Christian world. The Roman poet Catullus in 54 B.C. has described in his poem "The Marriage of Pelius and Thetis" the spinning action in which wool is converted into thread. In this action the wool on the distaff, which was originally made of wood, is twisted into thread as the distaff rotates. As part of this process some wool fibres stand out from the thread, due to electric attraction between the rubbed distaff and particles of the wool. These fibres were bitten off by the spinner's teeth. This description, written by a poet, may be the first written reference to electric attraction between a rubbed piece of wood and particles of wool. This attraction would be even greater if the wood were replaced by amber, as it might be in a wealthy home.

It was, however, many years later that the Arabian philosopher Averroes (1126-1198) described the attraction of rubbed amber for chaff as being due to "an occult virtue". This description took hold though it did not inform of a cause of the attraction. The phrase "occult virtue" was also used later by Thomas Aquinas (1225-1274). Because of his intellectual and spiritual stature, which lasted for many hundreds of years, this unhelpful description may have contributed to the lack of scientific progress in the immediately following centuries. But Jerome Fracastorio (1478-1553), a physician in Verona, near Venice, discovered that when diamond was rubbed, it too, like amber, would attract hairs and twigs.

The "occult virtue" of this electric attraction was paralleled by a similar, but much stronger effect known to Aristotle around 350 B.C. It was associated with a compound of iron, known to the ancient world as magnetite. This magnetite had the property of being able to attract both iron

1

and compounds of iron. A description of it is given in Aristotle's book entitled *Physica*. In recent years it has been found that the bacterium G-15 actually eats iron and converts it into magnetite. It is not known what proportion of the iron in the earth has been produced in this way.

The force of attraction of high quality magnetite is much greater than that of amber or diamond. It must have been of such a quality stone that Augustine (354-410 A.D.) wrote in his *City of God* as follows:

"I saw, I certainly saw, an iron ring snatched up and held aloft by a stone! And then it seemed as if the stone had given its own power to the ring which it had snatched up, and made it a joint property. For this first ring was applied to another ring, which it lifted aloft, and the second ring clung to the first, just as the first ring clung to the stone. In the same way a third ring was added, then a fourth, and in the end there was a kind of chain of rings hanging. The rings were not joined together internally by the interlinking of their circles, but adhering to each other from outside. Who could fail to be astonished at this property of the stone, which was not merely inherent in itself, but also passed on through so many objects suspended from it, and bound them together by invisible connections?"

It was also found experimentally, possibly even before the time of Aristotle, that the stone of magnetite, when freely suspended, always aligned itself substantially along a north-south axis. For much of the early history of magnetism, this property was of very great importance. It provided the only means of navigation over land or sea, for countries such as China or England.

The first step in scientific understanding, in place of the meaningless phrase "an occult virtue" was made by Alexander Nequam (1157-1215), who will be considered next. But this should not be taken to imply that there were no earlier users of magnetism. One such user was Hildegard of Bingen (1098-1179), whose general scholarship was outstanding. But since her work in magnetism was peripheral to that scholarship it will be dealt with after our consideration of Alexander Nequam. Thereafter contributors to our understanding of magnetism or electricity will be considered in chronological order. The two subjects of magnetism and electricity are closely linked, since any change of a magnetic field automatically produces an electric field, and conversely, as was discovered later.

Alexander Nequam (1157-1215)

Alexander Nequam was born in St Albans, on the same day as the boy who was to become King Richard I of England. His mother acted as a wet nurse to both boys. Alexander received his education at the Abbey School in St Albans. There his progress was so good that on leaving the school he was put in charge of a neighbouring school at Dunstable, some twenty miles away. From there he moved to the University of Paris, where at age twenty-three he became a Professor. In Paris, in addition to his teaching duties, he took time to study rhetoric, biblical criticism, canon law and medicine. Six years later he returned to his former position as Headmaster in Dunstable, and then applied to become a Benedictine monk. The story is told that the reply of the Abbott of St Albans to this application of Alexander Nequam's was given in Latin as, "Si bonus es venias, Si nequam, nequaquem." In English this translates as, "If you are good, come—if bad, by no means." This was a pun on his name, since "Nequam" means "bad" and "Nequaquam" is the Latin for "by no means". It is said that Alexander Nequam took offence at this, and instead became an Augustinian at Cirencester, where in 1213 he became Abbott.

Alexander Nequam also wrote a book entitled *The Nature of Things*. From this book it is clear that he believed the universe consisted of four elements, air, water, fire and earth—birds belonging to the air, fish to water, animals, vegetables and minerals to earth.

Scientific facts were believed to have a religious application. After the Day of Judgment, fire and water would disappear, so that only air and earth would remain. In connection with air he believed that the winds and storms associated with it are analogous to disturbances in the soul. Similarly spots on the moon should remind us of the spots in our own characters.

Nequam also believed that an agate worn on a person's body would make him amiable, eloquent and powerful. And that a lodestone, i.e. a piece of magnetite, placed on a woman's head, would make her confess her adulteries. Moreover, since a lodestone attracts iron he believed that the story of Mahomet's coffin being suspended in the air, within his mausoleum, could be explained by the attractive forces of powerful magnets placed in the walls of the mausoleum.

Another application of the lodestone mentioned by Nequam was its direction-finding property. When a magnetised needle stops rotating after being set in motion, it points along the north-south axis of the earth.

Abbess Hildegard of Bingen (1098-1179)

Earlier than the work of Alexander Nequam a supposed medical treatment for insanity was used by Abbess Hildegard of Bingen, near the River Rhine. Her treatment used a lodestone moistened with the patient's saliva, and drawn across his forehead. At the same time a prayer was repeated during the procedure.

The Abbess Hildegard has been described as the most talented woman of her age. It is said that she could tell the colour of a calf while it was still in its mother's womb. But apart from such an extra-sensory experience, her knowledge of herbs was such that she could treat more than a thousand medical conditions. In addition she was a poet, and a musician who could set her own poems to music. After she recovered from a persistent illness she described herself as "a feather on the breath of God". This was because she believed her continuing recovery was due to the Spirit of God moving her along gently. A three-volume work of visions she had was entitled *Know the ways of the Lord*, emphasising the life-giving power of the Holy Spirit.

On the strength of this assurance of the presence of the Spirit in her own life, Hildegard did not hesitate to express her views. When the Pope warned her against pride, she pointed out to him ongoings in the Papal household. Likewise to the Emperor Frederick Barbarossa in his dispute with the Pope she wrote: "My mystical insight shows me that you are behaving like a child—even worse, in fact like a fool." And this dispute ended with the Emperor having to prostate himself before the Pope. To King Henry II of England, who was later responsible for the murder of Thomas à Becket at Canterbury, she wrote. "Beloved son of God—call upon your Father, since willingly He stretches out His hand to help you. Now live for ever, and remain in eternal happiness."

One of her interesting sayings was that the gifts of God, i.e. spiritual gifts, do not grow old like old clothes. Instead they may grow into things of beauty. And although her use of magnetism did not add to our knowledge of it, it indicates the magical view which it held at that time. Twenty years after her death, it may have helped to bring forward the scientific view of Alexander Nequam mentioned above.

Historical Delay in Understanding of Magnetic and Electric Attractions

A period of about four hundred years elapsed between the publishing of Alexander Nequam's writings on the magnetic compass, and the first substantial book *De Magnete* written in Latin by William Gilbert in 1600. Such a delay was, in part, due to the original translations of Aristotle from Greek into Latin, not being of the quality necessary for their complete understanding. As a result, Pope Gregory IX in 1260 arranged for new translations to be made.

The views of Aristotle came to be gradually accepted as a result of these better translations.

Three scholars in Europe contributed to this acceptance. They were Robert Grosseteste (1168-1263) in Oxford, Albert Magnus (Albert the Great, c.1200-1280) in Cologne, and his chief disciple Thomas Aquinas (1225-1274) in Cologne and Paris. Thomas is believed to have requested the translations. Each of these scholars will now be considered in turn.

Robert Grosseteste (1168-1263)

To Robert Grosseteste in Oxford, we owe the distinction between an explanation based on logical reasoning from an experimental result, on the one hand, and an explanation based on the premise of Divine Providence on the other hand. When the first of these is used it may be possible to make a deduction leading to new facts.

Grosseteste was wonderfully able in mathematics, logic, language, and Biblical scholarship.

He became the first Chancellor of Oxford University, and was also a lecturer to the Franciscan House in Oxford. The Franciscans were begging friars set up by St Francis of Assisi in 1210, with vows of poverty. They had a care for the religious condition of the people. Because of the colour of the cloak they wore, they were commonly called Grey Friars. And they are not to be confused with the Black Friars, who were set up in 1215 in Toulouse by Dominic de Guzman (1171-1221). The emphasis of the Black Friars lay in education and learning, but both these Orders produced real scholars. It has been said that St Francis never condemned learning for itself, but had no desire to see it developed in the Franciscan Order. In his eyes it was not in itself an evil, but its pursuit appeared to him both unnecessary and dangerous. Unnecessary because a man may save his soul and also win others to save theirs, without learning. Dangerous, because it is an endless source of pride. As a simple means of differentiating the two Orders it has been said, "As poverty characterises Francis, so zeal for souls characterises Dominic."

Albertus Magnus (c 1200-1280)

Albertus Magnus (i.e. Albert the Great) joined the Dominican Order in Padua in 1223. After teaching at several convents, including Cologne where he was later to become famous, he went to the University of Paris. There he graduated as Master in Theology, and publicly taught the doctrines of Aristotle, despite a prohibition from the Church. He has been called Doctor Universalis because of the breadth of his scholarship. In1249 he became Rector at Cologne, with the specified duty of organising the University there. Albert was the only mediaeval philosopher to make commentaries on all the works of Aristotle, and he enriched science by his own observations in different branches of nature.

In theology the distinction belongs to him for resolving the difficulty between faith and reason, by basing himself, like Augustine, on two certainties. The first is the certainty of revealed religion. The second is the certainty of the faith of his own personal experience.

Both Albert the Great and Thomas Aquinas did not regard Aristotle as an absolute authority, but simply as a guide to reason. In particular, again in theology, their religious beliefs differed from those of Aristotle, again on two grounds. First they believed that the human soul was immortal, which was part of their Christian heritage, of which Aristotle could have no concept. Secondly Aristotle believed there was always motion in the world, as well as God. In contrast they believed that nothing except God alone has always existed.

Thomas Aguinas (1225-1274)

As an indication of the reverence in which the Roman Catholic Church held Thomas Aquinas, it is sufficient to say that in the Sessions of the Council of Trent (1545-1563) his *Summa Theologica* was placed on the altar along with the Sacred Scriptures and the Papal Decrees.

Thomas Aquinas was a scholastic giant. He brought together his knowledge from Arabic and Byzantine sources, and then related them to Christianity. Born in the castle Rocco Secca, near Naples, where his father was a Count, he was sent at age five to the Benedictine Abbey at Monte Cassino. This was on the border between the territory of the King of Naples and that of the Pope. When he was aged fourteen, war broke out between these two rulers and so he had to be sent home. He then attended the University of Naples, founded by the King of Naples as a counterbalance to the Papal University of Bologna. There he studied the liberal Arts of grammar, logic, rhetoric, geometry, music and astronomy. It was also possible for him to study Aristotle in Naples, study which was forbidden at that time in Paris.

As part of his religious training there, Thomas developed the practice of writing poetry, as in:

"Therefore we, before him bending
This great Sacrament revere.
Types and shadows have their ending
For the newer rite is here."

The long term acceptance of his writings is shown by the inclusion in *Hymns Ancient and Modern*, No.312, and also in *Church Hymnary* 2, No 319, of which the first verse of the hymn is:

"Thee we adore, O hidden Saviour, Thee
Who in Thy sacrament does deign to be;
Both flesh and Spirit at Thy presence fail,
Yet here Thy presence we devoutly hail."

At age nineteen Thomas Aquinas became a Dominican Friar, much to the displeasure of his family. This displeasure was probably due to the mendicant nature of the Dominican's life, in contrast to the wealth and learning associated with the Benedictine Order. The Benedictine Order was introduced to England by St Augustine of Canterbury about 6OO A. D., and there was a Benedictine Abbey School at Fort Augustus until a few years ago. In comparison, the Dominican Order appeared in Britain around 1221 A.D. But in Paris the Dominicans founded a study house in 1217, and by 1230 had obtained control of two of the twelve Chairs of Theology in the University of Paris.

It has previously been said in the section dealing with Robert Grosseteste that the emphasis in the work of the Dominicans lay in education and training. This is understandable since Dominic himself was a natural scholar who edited his notebooks carefully. But despite this he sold his books to help the poor in a time of great famine. Also, long before he founded the Dominican Order in 1215, he came across the Albigentian movement in the south of France. The Albigentians differed in doctrine from the traditional church, believing that man's aim should be the deliverance of man's soul from his body. Suicide by starvation was considered meritorious. Pope Innocent III launched a Crusade against them in 1209 A.D., during which twenty thousand people were slaughtered by the Crusaders. Later thousands more were killed on both sides before peace was restored in 1220 A.D. But before that happened, Dominic was appointed by the Bishop of Toulouse as Preacher to his Diocese. It was known that Dominic had experience of converting an Albigentian innkeeper by spending the whole night in debate with him, and bringing him back to orthodoxy. Dominic believed that heretics could best be won over by evangelical poverty, deep learning and zeal for souls.

The manner of life for these Black Friars was austere. They maintained a perpetual abstinence from meat, long fasts and silence. So great was the family displeasure that Thomas Aquinas chose to become a Dominican Friar that he was kidnapped for a year, to try to persuade him to change his mind. During his imprisonment an attractive woman was let in to his cell, but Thomas seized a brand from the fire and drove her out. This may explain his later habit of avoiding women.

The Dominican Order was very missionary minded, with centres in Europe, Persia, India, China and North Africa. Another feature of the movement was their involvement with the Universities. Dominic's policy was to have houses in all the famous University towns, Paris, Bologna and Oxford. Among their own teachers the most famous were Albert the Great and Thomas Aquinas. Nevertheless they were not immune to political interference. Thus in 1559 all the Dominican Houses in England were suppressed for about one hundred years. But they produced four Popes, and because of their emphasis on teaching the Catholic Faith, the Dominicans were entrusted with the conduct of the Inquisition.

From 1248-1252 Thomas Aquinas went to Cologne to study under Albert the Great. When he went there he was aged twenty-three. Under the tutelage of Albert, Thomas learned to appreciate Aristotle. Since Thomas never learned Greek it must have been through a Latin translation. This may account for the belief that he petitioned Pope Gregory IX for a more accurate translation of Aristotle.

Albert the Great was so famous that he attracted students to Cologne. Initially some of these students named Thomas "The Dumb Ox". Thomas was big physically and did not speak when speaking was not required. But it did not take long for the students to appreciate the quality of the notes which Thomas made from Albert's lectures.

After four years in Cologne, Thomas was sent by Albert to Paris, both to study for a Bachelor's degree, and to begin the theological course leading to a Master's degree. His initial task included lecturing on the Sentences of Peter Lombard. These comprised a collection of patristic texts on the Trinity, Creation, Christ and the Virtues, and the Sacraments. His treatment of these has been described in the following way:

> "God graced his teaching so abundantly that it began to make a wonderful impression on the students—for it all seemed so novel—no one who heard him could doubt that his mind was full of a new light from God."

So much for Thomas's early contribution to theology. But what was his contribution to Magnetism? In his commentary on Aristotle's *Physics*, it appears that he accepts what is our first written reference to the subject of magnetism. As one example Thomas writes of the pulling action of a magnet in the following way: "A thing pulls because it moves something towards itself, by altering it in some manner, from which alteration it happens that the thing altered is moved with regard to place. In this way a magnet is said to pull iron. It imparts some quality to iron through which the iron is moved toward the magnet."

Thomas then goes on to say three things about a magnet, all of which are partially true. But they suggest he did not carry out experiments himself. The three things he believed to be true were first that a magnet only pulls iron when it is near. Secondly if a magnet is "greased with other things", it will not attract iron. And thirdly if a magnet is small, the iron must first be rubbed with the magnet.

The work of Thomas Aquinas, however, which is best known to the public is *Summa Contra Gentiles*. This was written at the Papal Court between 1259 and 1265, when he was aged from thirty-four to forty. The purpose of the work was described in 1313 as follows:

> "St Raymond of Penafort, strongly desiring the conversion of unbelievers, asked an outstanding Doctor of Sacred Scriptures, a Master in Theology, Brother Thomas of Aquino of the same Order, who among all Clerics of the world was considered in philosophy to be, next to Brother Albert, the greatest, to compose a work against the errors of unbelievers, by which both the cloud of darkness might be dispelled and the teaching of the true Sun might be made manifest to those who refuse to believe. The renowned Master accomplished what the humility of so great a Father asked, and composed a work called the *Summa Contra Gentiles*, held to be without equal in its field."

The Arabic interpretation of Aristotle's writings had been unacceptable to Christian thinkers.

So Thomas Aquinas turned his massive intellect to putting a Christian interpretation on them. The result was *Summa Contra Gentiles*. It comprises three hundred thousand words in four volumes, and was designed to show, by natural reason, to which all men are forced to submit, a limited number of truths about God. But beginning with God it is not unreasonable to postulate that the Divine intellect far surpasses the angelic intellect, which in turn far surpasses the human intellect. Acceptance of this principle has the effect of curbing human presumption, which is the source of much human error.

Many quotations could be given from, or about, the writings of Thomas Aquinas. The following two will give a flavour:

> "The beatitude of spiritual creatures is the central purpose of creation. Thus the physical universe is anthropocentric in a way that remained unknown to the Greeks and Arabs."

"Thomas Aquinas's mind was entirely devoted to the intellectual task of being a disciple and a student of the Truth contained in the Christian revelation."

But his intellectual devotion was not one that led him to an ivory-tower existence. For example, he became a strike breaker in his first year in Paris when the University attempted to suppress one of the two Dominican Chairs. Because this resulted in the Professors going on strike, Thomas gave lectures in their place. He also declined the positions of Abbot of Monte Cassino and Archbishop of Naples. Moreover, in one of his earlier works, *Sententias* II, dist.44, he stated, "The Pope, in virtue of his canonical office, is the Spiritual Head of the Church, and nothing else. Every other political or worldly accretion is a historical accident."

ROGER BACON (1214-1292)

Whereas Thomas Aquinas is universally respected for his scholarship, associated with the Dominican Order of Friars, the contribution to science and philosophy of Roger Bacon is of a different kind. It will be recalled that Thomas Aquinas was primarily a thinker, who did not think it essential to check the results of his thinking against experiment. Although Thomas's tutor and mentor, Albert the Great, valued the results of experiment, he did not succeed in transferring this appreciation to Thomas. In contrast, Roger Bacon, born in Somerset, and trained initially in Oxford under Robert Grosseteste, was what might be called a natural experimenter, in addition to also being a philosopher.

After graduating M.A. in Oxford, Roger Bacon taught there for a time, before transferring to the University of Paris, where he lectured from approximately 1237-1245, i.e. when he was aged 23 to 31. The subjects of his lectures there followed Aristotle, both scientifically and also metaphysically. Thereafter his concentration on the scientific side seems to have begun around 1247 when he returned to Oxford, and this lasted for twenty years.

During the whole of his life he had the reputation of being an unconventional scholar. He could be outspoken, not hesitating to attack his contemporaries. Possibly as a result of this he was accused by them of "certain suspected novelties". This may have been the result of religious jealousy by the Franciscans, of which Roger had become a Friar in 1256. Four years later in 1260 a Franciscan statute was passed forbidding any publication without approval. The only way this statute could be circumvented was by a Papal Order, and Pope Clement IV issued such an Order in1265. As a result Roger was able to return to his writing, producing three publications, *Opus Maius*, *Opus Minus*, and *Opus Tertium* over the next two years. These books constitute an Encyclopaedia of Science.

But Roger's main interests were in Alchemy and Optics. He was one of the first people to use experiments in Alchemy, and his investigations into refraction of light led eventually to the development of spectacles. His interest in gunpowder contrasted with his attacking the corrupt text of the Latin Vulgate, and the production of a Hebrew and a Greek grammar.

But Roger Bacon will also be remembered for his relationship with Peter Peregrinus, i.e. Peter the Pilgrim, a title given to those who engaged in a Crusade under the aegis of the Pope. Bacon wrote an extraordinarily high commendation about Peter the Pilgrim, although the only surviving record of Peter's writings came in 1269, two years after the commendation was written.

The substance of Peter Peregrinus's writing to a fellow soldier is magnetite, or the lodestone. He points out its usefulness to a traveller who has gone astray. Recognition of a good lodestone depends on its colour, homogeneity and weight. Any lodestone will possess North and South poles, with attraction existing between the North pole of one lodestone and the South pole of any other. A long piece of iron which has touched a lodestone will align itself with the North-South axis of the earth.

This Epistle of Peter Peregrinus was written during the siege of Lucera. This town was supporting the Saracens against Pope Clement IV who had conferred on the expedition all the privileges of a Crusade. Lucera fell in 1269, the date of Peter's letter.

Duns Scotus (1265-1308)

Although Thomas Aquinas was canonised in1323, and declared by Pope Pius V in 1567 to be "The Angelic Doctor", it was not until 1993 that Duns Scotus was beatified. However he was given the title of "The Subtle Doctor" for his subtle merging of different views, after1304, when he was Professor in the Theological Schools in Paris.

Duns Scotus was born in Duns, Berwickshire. He was buried in Cologne, where his tomb bears the inscription in Latin: "Scotia brought me forth, England sustained me, France taught me, Cologne holds me." It is recorded that when he was no more than a boy aged twelve, but had been grounded in grammar, he was taken by two Scottish Minorite Friars (The Grey Friars) first to a Franciscan convent at Dumfries, and in 1290 to Oxford, for there was no University in Scotland at that time. St Andrews was not founded until 1411. In 1301 he was appointed Professor of Divinity at Oxford, and such was his reputation that students came from all parts to listen to his lectures. Soon afterwards he was called to teach at the University of Paris, but was not long there when he was exiled for refusing to support Philip IV, the King of France, in a quarrel with Pope Boniface VIII over the taxation of church property. After an exile of a year he was allowed to return to Paris, but was then sent to Cologne in 1307, where he died a year later.

Theological Views of Duns Scotus

Scotus took the view that the primary concern of Theology is God, whereas Philosophy treats God only as the first cause of things. Regarding Theology as a science, Scotus saw it from the practical point of view as being concerned with saving souls through Revelation. He argued that a person may know by faith, with absolute certainty, that the human soul both exists and is immortal. In contrast to this, reason can only postulate the existence of such a soul.

It must be said that Scotus laid a great emphasis on Love. For himself he laid down three requirements. The first was that he must love God. The second was that he must love other people as people who should also love God. The third was that he must love himself as someone who should love God. This emphasis on Love had a life changing effect on the poet G. M. Hopkins (1844-1889). Hopkins had believed that love for even one friend resulted in neglecting one's primary duty of Love to God—his love for the specific was being antagonistic to the Divine. But Hopkins's study of Scotus convinced him that the aesthetic and religious experiences became one. This gave him the warrant to be able to love with his senses. Hopkins then went on to express his gratitude to Scotus in the sonnet "Duns Scotus's Oxford", from which the following lines are taken:

"Yet ah! This air I gather and release
He lived on; these weeds and waters, these walls are what
He haunted who of all men most sways my spirits to peace;
Of reality the rarest-veined traveller; a not
Rivalled insight, be rival Italy or Greece;
Who fired France for Mary without spot."

The final line quoted refers to the Virgin Mary, whose Immaculate Conception, i.e. the belief that she was conceived without Original Sin, was defended by Scotus. At the time the general belief was that it was appropriate, but the problem was how to accommodate the belief that only with Christ's death would the stain of Original Sin be removed. Scotus, citing Anselm's principle "God could do it, it was appropriate, therefore He did it", Scotus argued that Mary was in need of redemption like everybody else. But through the merits of Christ's crucifixion, given in advance, she was conceived without the stain of Original Sin. This view was defined as a dogma of the Roman Catholic Church by Pope Pius IX in 1854.

Another of Duns Scotus's contributions, this time in opposition to that of Thomas Aqulnas, was "his assertion that the individual quality of a thing, which he called its 'Thisness', is the final perfection of any creature." He asserted that the individual is immediately knowable by the intellect in union with the senses. In particular it is true to say that for man, the Will as the active principle in Thisness, has priority over the Intellect. This is in keeping with the saying "Intellect proposes, but Will disposes".

A small number of quotations from Scotus follows:

"Being elected by God implies also being blessed by God. The one cannot be true unless the other is also true."

"The relation between God and man is a relation of Love. The Love of God calls for *love from* man."

"God is both free and good. Since man was made in God's Image, man's freedom and goodness reflect God's freedom and goodness."

Duns Scotus was, of course, one of his time, and was not immune to the problem of the Jews. He believed in the necessity of preserving a Jewish remnant until the return of the Messiah. His suggestion for doing this was the deportation of a certain number of them to a distant land, and keeping them there at Christianity's expense to the end of days.

Predecessors of Duns Scotus

It would be wrong to think that Duns Scotus's ability existed in isolation from his fellow countrymen. One of his predecessors was Richard Scot (c.1123-1173) who influenced greatly the theological stance later adopted by the Grey Friars. He too placed Love at the centre of his theological system. But he was probably influenced by the Anglo-French Anselm of Canterbury (c.1033-1109) who after his enthronement as Archbishop, to which he was implacably opposed, hastened back to his monastery in Gaul to go on with his interrupted studies.

Another predecessor was Michael Scot (died c.1236). He translated Aristotle's *Physics* from Arabic into Latin with a commentary by Averroes (1126-1198), the famous Arabian philosopher of Spanish origin. At one time Michael Scot was Astrologer to the King of Sicily who founded the University of Naples in1224.

Followers of Duns Scotus

In England, William of Occam, born in Surrey, became a Grey Friar in 1300. His attitude to poverty, after being educated at Oxford and Paris, was such that it led to conflict with Pope John XXII. As a result he was imprisoned at Avignon, but escaped and was received by the Emperor Louis of Bavaria, who was also in conflict with the Pope. After being excommunicated he spent the rest of his life at Louis's Court, but was appointed General of the Franciscan Order in 1342.

Occam asserted that Aristotle believed that a magnet must be in contact with a body before it can be moved. Against this Thomas Aquinas's commentary asserts only that it must be close to the body to be moved. Scotus saw no difficulty in believing in the theory of action at a distance, using the analogy of the moon exerting an influence on the tides, like the influence he said of lodestone on iron. His belief, however, was not universally accepted, with the consequence that a belief in the existence of an aether held for many generations.

Another follower was John Mair, who was born near Haddington c.1467. He became a student in Paris, and was later appointed to be Professor of Theology there. In 1518 he became Principal of Glasgow University, then moved to St Andrews, then back to Paris, where two of his students were Buchanan and Calvin, and finally back to St Andrews again. John Knox referred to him as an oracle on matters of religion, and he was the author of more than forty books.

The Lodestone's Greatest Discovery—America

Christopher Columbus (1451-1506)

The name Christopher means "carrying Christ", and Christopher Columbus took this meaning as his duty in life. Most people remember him rightly as the discoverer of America. But when he himself reached the island of Haiti in the West Indies, and discovered the presence of gold there, his first thought for its use was the recapture of Jerusalem from the Saracens. The Crusades, all eight of them, had ended in failure, following the capture of Acre by the Sultan of Egypt in 1291. But Columbus believed he was called to the task of recovering Jerusalem.

For forty years before the discovery of America in 1492, the King of Portugal had been responsible for sending ships south, even round the Horn of Africa, on journeys of discovery. But these did not reach India, because the crews refused to go further into the unknown. Columbus had a vision of reaching Japan and China by sailing west. Of course the same arguments of sailing into the unknown would still apply, but Columbus believed the distances were not impossible. In this belief he would have found support from 2 Esdras 6 v.42, where we read that on Day Three of Creation: "You ordered the waters to collect on a seventh part of the earth; the other six parts you made into dry land, and from it kept some to be sown and tilled for your service. Your Word went forth, and at once the work was done."

This quotation, of course, is from the Apocrypha, and not from the Old Testament. The fact stated is in error. But it is easy to imagine how a religious person, like Columbus, could have believed this error, against the current belief of the time that the sea distance to Japan and China might be impossibly large. In fact Columbus did not realise on his first voyage westwards that he had discovered a new continent. He had set out to discover what he called the Indies, which he did, and with it gold as well.

Although he reached his objective in 1492, Columbus had initially approached the King of Portugal in 1484 with a view to reaching what was called gold-roofed Japan. As an experienced mariner, Columbus asked for more than one ship. When this request was refused by Portugal, Columbus went to Spain the following year, but obtained the same answer. However he persisted and received some royal support until 1488. In the previous year Bartholomew Dias had rounded the Cape of Good Hope, but his crew refused to go on to India. By 1491 Columbus had three rejections from the King and Queen of Spain, but in January 1492 the Queen of Spain decided to support his application.

In the event three ships were provided, and they left from Palos on 3rd August 1492. Before leaving, every man and boy had to confess his sins, receive absolution and take communion. As far as navigation was concerned it appears from those who have studied Columbus's journal that he relied virtually entirely on Dead Reckoning. That is to say the calculation of the ship's position was obtained from the course steered by the compass, which used a magnetised needle, and the distance travelled was estimated from the speed multiplied by time. As the needles did not maintain their magnetism well, a lodestone was always available for remagnetisation. It appears that celestial navigation was not practised much by the particular crews involved in Columbus's voyage of discovery.

On Christmas Day 1492 one of the three ships ran aground on the island of Haiti, because of failure by the Duty Officer on watch. This resulted in the total loss of the ship, though most of the stores were saved. Columbus took the view that the shipwreck was the predestined will of God, because help from the local islanders brought to Columbus's notice the existence of many gold artefacts. These had been made from gold mined from a nearby gold mine, the existence of which Columbus had been totally unaware.

Geronimo Cardano (1501-1576)

The Milanese born Cardano was a mathematician, philosopher and physician. He is perhaps best known for his contribution to solving algebraic cubic equations. But he was also a physicist who was aware of the use of amber in the manufacture of printing ink and incense. In addition he knew that when burned it was believed to keep the plague away. Consequently there was a market for it when it was carved into rings and rosaries.

His contributions to physics lay in observing the differences between electric and magnetic attractions. These he noted were:

(1) Amber attracts everything which is light in weight, whereas a lodestone attracts iron only.
(2) Amber does not draw chaff when something is interposed, whereas a lodestone does.
(3) Amber does not attract at its ends, whereas a lodestone does.
(4) The attraction of amber is increased by heat and friction, whereas the attraction of a lodestone may be increased by cleaning it.

Cardano also made use of his observational skills in his medical work. He is known to have treated the Pope, by whom he was later given a pension. He also treated Archbishop Hamilton of St Andrews, who had been ill for ten years, until Cardano diagnosed his allergic reaction to feathers in his feather bed.

In summary the three major advances in magnetism and electricity which were made up to the middle of the sixteenth century, were made firstly by Alexander Nequam, a part-time priest, secondly by Peter Peregrinus, a soldier, and thirdly by Geranimo Cardano, a mathematician, physicist and physician. This practice of advances being made, mainly or partly, by those working outside the Church, continued in the case of Robert Norman, who will be considered next in Chapter Two.

CHAPTER TWO

DOMINANCE OF EXPERIMENTAL RESULTS

(c.1550 A.D. – c.1740 A.D.)

ROBERT NORMAN. (His only known dates are those of his published books in 1581 and 1590.)

Robert Norman of Radcliffe, England, spent eighteen to twenty years at sea before becoming a full-time Instrument Maker of Magnetic Compasses. He wrote a book in 1581 called *The New Attractive*. This begins with a poem entitled "The Magnetite Challenge", in which he compares the beauty of jewels with the usefulness of lodestone, otherwise called magnetite. From this poem the following small number of lines is selected:

The Magnetite Challenge

Give place ye glittering sparkes
Ye glittering diamonds bright
Ye rubies red, and sapphires brave,
Wherein ye most delight.

Magnes, the loadstone I,
Your painted sheaths defy,
Without my help in Indian seas
The best of you might lie.

Robert Norman in his seagoing experience noted the different variations in angle over the earth's surface between true north and magnetic north. He also knew that magnetite's ability to point in a particular direction could be transferred to iron, together with its ability to attract iron. And in turn this iron could attract more iron, as described so enthusiastically by Augustine in his *City of God*. Robert Norman thinks rather similarly to Augustine, writing:

> "True it is, that God is mighty and marvellous in all His works, yet He does not allow us to say more than truth of them. And truly His power is as greatly shown in the magnetite as in any stone that He has created. And who so shall go about curiously to seek out the efficient cause of His properties, I suppose the longer he seeketh, the more he shall marvel, and yet never the nearer his purpose."

It is readily understandable that in the history of magnetism reasons should have been sought for the directional properties of magnetite. Some considered the existence of a source point in the heavens. Others considered this source point to be in the earth, near the North Pole, imagining the existence there of great rocks of magnetite. But Norman rejected this idea, since he knew that terrestrial sources of magnetite did not significantly affect compass readings more than a quarter mile from them.

It was at this point that Norman made his major discovery from his constructional knowledge of magnetic compasses. Having been asked to make a compass with a needle of length five inches he carefully balanced the needle before magnetising it. After the magnetisation the north pole of the needle was found to dip. To measure the angle of dip the needle was supported on a horizontal pivot, and the dip angle was found to be about 72 degrees to the horizontal. The source of attraction to its north pole thus exists along a straight line which proceeds from the midpoint of the needle into the earth, but at some unknown distance from it. This direction-finding property of the needle had been derived from the magnetite, but to quote his own words, "As to how it is engendered I am no more able to satisfy you than if you should ask me how the celestial spheres are moved: but God in his omnipotent providence has so appointed it to be."

Norman's interest in establishing and using experimental results was based on his concern for safety at sea. Following the publication in1581 of his book *The New Attractive*, he turned his attention to translating a Dutch manual, *The Safety of Saylors* (1590). This contains the "Course, Distances, Soundings, Floudes and Ebbes with the marks for the entering of sundry harbours in England, France, Spain, Ireland, with other necessary rules of common navigation."

As in *The New Attractive* he begins *The Safety of Saylors* with a poem:

> *In Commendation of The Painful Seamen*
>
> Whoso in surging seas, his season will consume
> And means thereof to make his only trade to live:
> That man must surely know the shifting Sun and Moon
> For trying of his Tides, how they do take and give.
>
> If pilot's painful toil be lifted then aloft
> For using of his Art according to his kind;
> What fame is due to them that first this Art out sought,
> And first instructions gave to them that were but blind.

This poem is followed by instructions: "How to sail from Amsterdam going out to sea", and "How to sail to Amsterdam coming out of the sea", and many other details for other situations. It also provides information about tides along the coast of England and Ireland, sands at the mouth of the Thames, tides off the coast of Flanders, and brief notes of St George's Channel. In all, it is a very practical handbook.

But in the year preceding its publication, in 1589, a very different, but significant book, entitled *Natural Magic*, was published by John Baptista Porta, a Napolese physicist.

JOHN BAPTISTA PORTA (c.1540-1615)

John Baptista Porta has been described as a child prodigy. He was born of a noble family in Naples, and it is said that he produced his first book on "Natural Magic" at the age of sixteen. Later in life he expanded it in both substance and scope into a substantial volume covering twenty areas of knowledge. The volume with the most lasting value dealt exclusively with Magnetism. The breadth of his knowledge and interests, however, is shown by the other books which include the reproduction of animals, the transmutation of metals, statics and the preparation of perfumes. In addition he wrote comedies, and was the founder of the first of all learned societies, the Academia Secretorum Naturae. In this society it was an essential condition of membership to demonstrate research ability in either medicine or philosophy. But perhaps because its novelty frightened either the general populace or the entrenched powers that be, Porta was summoned by the Pope in Rome, and the Society was disbanded at that time. Later a related Society was founded in Rome, of which Porta became a member.

Della Porta wrote on many subjects as is shown by his twenty volume book on "Natural Magic". Volume Seven, of length twenty-six pages, is the one dealing exclusively with Magnetism. How much of this is Porta's own work it is impossible to say. He acknowledges help from Fra Paolo Sarpi (1552-1623) of whom he said, "I do not blush, but consider myself honoured to confess, that many things concerning magnetic phenomena I have learned from Paolo, a true ornament of light, not only of Venice, but of Italy and the whole world."

But certainly Della Porta did make some contributions of his own, marred sometimes by his over enthusiasm. For example, he was so sure that magnetism was able to provide information on longitude that he put this statement into the Preface of his book. But later it was found that it was useful over short longitudinal distances only. Likewise, he conceived the idea of coupling between magnets as being sufficient to permit communication with people in prison, if magnets were available at both ends of the communication link. Similarly, he believed, like Peter Peregrinus, that perpetual motion could be achieved through magnetism. But some of his points were correct. Thus he asserted that heating destroys magnetism. Moreover he pointed out that the strength of a magnet could be measured by weighing the increase in weight produced when the magnet operates on an iron object placed on one side of a chemical balance. A table of wood, stone or any metal except iron does not affect the operation of a magnet. Sailors prefer steel needles in a mariner's compass because its magnetism lasts for a very long time. Magnets may be used in mines and tunnels as well as in Mariners' compasses. The existence of a field of force surrounding a magnet, diminishing with increasing distance from its source, was analogous to the similar diminishing of light from a candle with increasing distance.

The reference to Della Porta's acknowledgement of the great help he received from Fra Paolo Sarpi (1552-1623) will now be expanded, by including an independent section on Sarpi.

Fra Paolo Sarpi (1552-1623)

It may be true to say that there are few people in the world about whom such contradictory views are held as about Fra Paolo Sarpi. His opposition to Pope Paul V from the time of his Papal appointment in 1605 has had him described by a Catholic historian as the most anti-papal churchman after Martin Luther. Contrary views are, however, legion. Thus Edward Gibbon called Sarpi, "The incomparable Historian of the Council of Trent."

Lord Macaulay says of him, "What he did he did better than anyone else."

Galileo called him "My father and my master" and also said of him, "No man in Europe surpasses Master Paolo Sarpi in his knowledge of the science of mathematics."

John Donne, when Dean of St Paul's, had a portrait of Sarpi, a fellow enemy to the Jesuits, in his parlour.

Sir Henry Wooton, the first resident English Ambassador to the Venetian Republic from 1604, called him "the most deep and general scholar of the world" and "a true Protestant in a monk's habit".

John Milton refers to him as "the great and learned Padre Paolo".

Gilbert Burnet, formerly Professor of Divinity at Glasgow (1669), author of *The History of the Reformation in England*, and later Bishop of Salisbury called him "a man equally eminent for vast learning and a most consummated prudence, and was at once one of the greatest Divines, and of the wisest men of his age".

Against this it has to be said that in Whittaker's scholarly book *History of the Theories of Aether and Electricity* there is not a single reference to Sarpi or his work. One reason for this may be associated with Whittaker's conversion to the Church of Rome, which may have coloured his judgment.

Pietro Sarpi and the Servants of Mary

Pietro Sarpi was born in Venice in1552, the year in which Francis Xavier the Jesuit missionary died. His father had the reputation of being of a choleric temper, but his mother was pious, the sister of a clerical uncle who conducted a stadium or school. Pietro was placed in this stadium when his father died, and showed himself to be exceptionally gifted—so much so that he was soon transferred to a Servite Ministry nearby.

The Servites or "Servants of Mary" was a religious Order established in 1233, when, quoting from the *Catholic Encyclopaedia*, "the Blessed Virgin appeared to seven noble Florentines in church and bade them leave the world, and live for God alone. On a later occasion she addressed to them the following words: I have chosen you to be my first Servants, and under this name you are to till my Son's vineyard. Here too is the habit which you are to wear; its dark colour will recall the pangs which I suffered on the day when I stood by the Cross of my only Son. Take also the Rule of Augustine, and may you, bearing the title of my Servants obtain the palm of everlasting life". The Order for women was expanded into England in1850.

Paolo Sarpi entered the Servite Order in Venice at age fourteen. At that time his tutor was the Friar Capelli, who was an expert on the writings of Duns Scotus. But it became clear that Sarpi was able to fault some of Duns Scotus's arguments. He was gifted with an excellent memory and a studious disposition, not taking part in the normal activities of youth. In this sense of avoiding such activities he was a forerunner of Henry Scougal, who also shared a studious disposition, and became Professor of Divinity at Aberdeen a century later.

Sarpi, against the wishes of both his mother and his uncle, chose to become a Servite Friar. Such was his ability that at age eighteen he engaged in a theological disputation at Mantua, publicly defending three hundred and eighteen theological and philosophical theses. This was done in the presence of the reigning Duke Giugliielmo III, who thereupon appointed him Court Theologian, with the local Bishop also appointing him Professor of Positive Theology. In addition his superiors awarded him six crowns annually, to buy books.

For four years Sarpi remained at Mantua, until 1574, making there the acquaintance of Camillo Glivo, who had been a Secretary at the Council of Trent which had been held from 1545-1563. He also worked under Carlo Borowmen, a Counter-Reformation Bishop in Milan, who invented the Confession Box, and who attempted to make men and women worship in separate churches. There was therefore little common ground between him and Sarpi.

In 1574 Sarpi returned to Venice to teach philosophy at the Servite Monastery. This was not an esoteric subject divorced from real life, because while he was there in 1576, plagues in Milan and Venice resulted in the deaths of tens of thousands of people. Sarpi continued his scholarly work in

Venice, and two years later was awarded the degree of Doctor of Theology by the University of Padua, becoming the youngest recipient ever. He was also elected Prior of the Venetian Province of his Order. By age thirty five (in 1587) he was the second highest official in the Servite Order. During this time in Venice he also developed a friendship with Arnaud du Ferrier, the French Ambassador to Venice, who had represented the King of France towards the end of the Council of Trent, and who has been described as being a secret convert to Calvinism.

Pietro Sarpi and Anatomy

Between 1582 and 1585 it is believed that Sarpi devoted himself to the study of Anatomy. From his examination of the specific gravity of blood "He concluded that there must be a mechanism in the veins by which the blood could be suspended, and its flow regulated so as to avoid dilatation and congestion, as in varicose veins. Having arrived at this conclusion a careful examination with the dissecting knife and microscope confirmed this theory". He at once made this known to Girolamo Fabrizi d'Aquependente (1537-1619), who was Professor of Anatomy at the University of Padua from 1562, and who, in turn, announced it in turn to his students as a communication from Sarpi, "the oracle of the century". William Harvey from England was a member of Fabrizi's class at that time.

In support of the view that Sarpi also discovered the circulation of the blood, Weesseling, holder of a Chair of Anatomy at Padua, at the same time, says he saw in the hands of Fra Fulgentia his Friar companion of a lifetime, a paper in the writing of Sarpi on the subject of blood flow. He testified that "the circulation of the blood demonstrated by Harvey had first been demonstrated by Fra Paolo Sarpi of Venice, from whom also Acquipendente learned the other discovery of the valves of the veins. Further, Bartholin (1614-1680), the Dutch Anatomist who was a friend of Harvey, and was the first to give experimental proof of the circulation of the blood, believed that to Fra Paolo Sarpi belonged the merit of its discovery. Harvey's book *Exercitatio Anatomica de Motu Cordis et Sanguinis* was not published until 1628, five years after the death of Fra Paolo Sarpi.

Pietro Sarpi and Magnetism

With regard to Magnetism, Alexander Robertson D.D. has written in his biography of Sarpi, "Of magnetism, Fra Paolo Sarpi knew more perhaps than anyone of his time, and he quotes Fra Fulgenzio, his Friar companion, in the story of a great magnetician who came to Venice from beyond the mountains, and who had an interview with Fra Paolo in his cell." It is not known if this might have been Wiliam Gilbert. "He thought to enlighten him, but soon found Fra Paolo knew all about it and a good deal more." It is known that Sarpi corresponded with Gilbert (1540-1603), who did visit him in his cell at the Servite Monastery in Venice, where he found that Sarpi had arrived at the same conclusions he had in regard to the earth being a great magnet. Fra Fulgenzie tells us that Gilbert was amazed to find that in every one of his discoveries he had been anticipated by the Italian Friar.

Another biographer Griselini, after dwelling on Sarpi's profound knowledge of magnetism, concludes, "And now I declare that the writing of Sarpi which exists on this matter justifies me in saying that in Gilbert's book, there is nothing which has not been previously observed and experimented on by Sarpi." Gilbert himself bears witness to Fra Paolo's wonderful knowledge of magnetism, setting him above Della Porta, the Neapolitan Professor of Natural Philosophy.

Pietro Sarpi and the Council of Trent

In Sarpi's own writings which have not been destroyed by fire, his major work is regarded as *The History of the Council of Trent*. This was completed in 1616, and published in England under a pseudonym in 1619. Although Sarpi had not been born when the Council started, and was only eleven years old when it finished, he had access to both Camillo Glino and du Ferrier, both of whom had been present during parts of that Council.

Prior to the publication of that work, Sarpi had effectively entered politics as a result of his intellectual reputation in Venice. The theologian that he was, enabled him to respond to attempts by the Pope in Rome to obtain control over the Church in Venice. This struggle between Rome and Venice had been ongoing for some time, but came to a head after the accession to the Papacy of Pope Paul V in 1605. The immediate cause of the rupture was the decision by the State of Venice to try two ecclesiastics in a State Court, whom the Pope submitted should be tried in an Ecclesiastical Court. The Pope demanded the release of the two prisoners in December 1605, together with the repeal of certain Venetian laws regarding church property. Failure to meet these demands would result in the State of Venice being interdicted.

An interdict is the excommunication of an entire district or country. It forbad the celebration of the Mass, the administration of the Sacraments, any Christian burial and baptism of babies. In January 1606 the Venetian Senate refused to comply and named Fra Paolo Sarpi as their Consultant. The Senate declared the Interdict invalid and ordered their clergy to perform their religious duties which Rome had forbidden. One Venetian clergy said he would not say Mass on an important occasion. A representative from the Council enquired on the preceding day how he would act. He replied he would act next day as the Holy Spirit prompted him. When this reply was received by the Council they sent a message back to say that the Holy Spirit had instructed them to say that if he failed to perform his expected duty he would be hanged at the door of his own Church. He said the Mass. The Jesuits were expelled for refusing to say Mass, though they were willing to receive confessions.

A compromise between the Papacy and the State of Venice was later reached, partly through the intervention of a French Ambassador. The two clerical prisoners were handed over to the Church Courts, but Venice did not surrender its right to judge such offenders. Sarpi was excommunicated in January 1607 for refusing to go to Rome to justify his conduct, but the Interdict was lifted in April 1607. Nevertheless, Sarpi was subjected to a knife attack in Venice in October 1607, from which he was able to recover by June 1608, after Acquapendente had been brought from the University of Padua to assist his recovery. But somewhat parallel magnetic work to that of Sarpi had been going on in England, mainly due to William Gilbert.

Wiliam Gilbert (1544-1603)

William Gilbert was one of the Court Physicians to Queen Elizabeth of England, and also to her successor James VI of Scotland and I of England. His father was a lawyer who became Recorder of Colchester. He had five children by his first wife, and of these William was the oldest. At the age of fourteen he matriculated at St John's College in Cambridge, where he took his B.A. degree in 1561 and his M.A. in 1564. He was appointed a Mathematics Examiner in 1565 and 1566, and Bursar in 1569 and 1570. He also took the M.D. degree in 1569, and this was to be the basis of his later professional life in Medicine.

But William Gilbert is chiefly remembered for his scientific rather than his professional life. The basis of this probably lay in his teaching duties at St John's College, where he is said to have recorded weather patterns in association with astrological data, in accordance with the Galen system of a doctor's training at that time. This looked for similarities between the symptoms of a patient and characteristics associated with different planets. It was in London that he chose to practise as a doctor, and it was there that he developed his skill in working as a member of a group. Medically this was associated with the College of Physicians, but in later life it was transferred to skills involved in working with both mathematicians and navigators. He never married. In the College of Physicians he rose to become President of the College in 1600, the same year in which he published his world famous book of some three hundred and seventy pages on the magnet, *De Magnete*.

At its very beginning *De Magnete* is unusual in not being dedicated to a patron, which was usually the case when the author was financially supported in this way. But it must be remembered that Gilbert was a successful doctor. Nevertheless the experiments which are recorded in the book must have cost a great deal of money. It has been suggested that Queen Elizabeth herself may have contributed towards this cost, although it will be borne in mind that Gilbert's appointment as a Court Physician came at the end of his experimental work. In place of such a dedicatory introduction, *De Magnete* has an address by Edward Wright, who was a Cambridge trained mathematician, with navigation experience obtained on an expedition to the Azores in 1596. This was followed by a period as a lecturer for the East India Company, and his writing of a book entitled *Certain Errors in Navigation*, published in 1599 and dedicated in its second edition of 1609 to Prince Henry, son of King James VI and I. But only two years later Prince Henry died, aged eighteen, and Edward Wright who had been made Tutor to the Prince, no longer had this job. He died in 1615. Wright himself believed that a magnetic method for finding latitude was sound, but the rate of change of the angle of dip was such that its accuracy in finding this latitude was not sufficient to be useful in practice.

Although the authorship of *De Magnete* is always ascribed to Willam Gilbert, it is undoubtedly true that the book contains some information which comes from Edward Wright's own work. But this is not to detract from the tour de force which the book was, and it was Gilbert who predominantly produced it. The contents of the book are divided into six sections:

Book 1: Writings of Ancient and Modern Authors Concerning the Lodestone

This gives a history of magnetism, where magnetic material in the form of lodestone is found, what the poles of the lodestone are, and how they are found, attraction between unlike poles, attraction of iron, and the assumption of north and south directions by a long piece of iron. In the last chapter of Book 1 Gilbert postulates that the earth is a giant lodestone with poles and an equator, and draws objects to itself. Here Gilbert has gone beyond saying that the earth is magnetic because it contains iron, saying that it is a giant lodestone. It may be that this revelation came to him from working with a small spherical lodestone which he called a terrella, and which he must have made to his specification. But he would have known that three hundred years before, it was a spherical lodestone which had also been used by Peter the Pilgrim in his early experiments on the magnet. There was, however, a fundamental distinction between the concepts of these two experimenters. Peter the Pilgrim regarded his terrella as a model of the spherical heavens, whereas Gilbert regarded the terrella which he was using, as a model of the spherical earth. The constituent parts of this earth he did not accept as being the Aristotelian elements of air, water and fire, but as a single element of earth.

Book 2: Of Magnetic Materials

In Book 2 a distinction is made between an amber, i.e. an electric effect, and an iron, i.e. a magnetic effect. Although the attractive effect of amber was well known, Gilbert carried out careful experiments to show that many other elements shared this property. Examples were gems, glass, antimony and sulphur. In addition they attracted materials other than chaff, wood, leaves, stones, even particles of metal. But unlike magnetic material, these electric materials needed to be rubbed to demonstrate the effect. And the effects depended on a dry atmosphere, and the absence of intervening material between the rubbed object and the attracted item. In contrast magnetic substances were able to show their effects through any substances other than iron.

To show whether a substance was electric or not, Gilbert devised an instrument in the form of a rotating needle, which he called a versorium, from the Latin "to whirl around". This consisted of a finely balanced horizontal metallic needle. If the needle turned when the rubbed material was brought close to it, this indicated that the material was electric. Such turning was always due to attraction. A similar measuring system was also used with magnets, but in this case the rotating needle had to be made with iron.

Gilbert also discovered that the effective strength given to an iron rod was greater than to an iron sphere, or cube, or iron in any other shape. With regard to a terrella he also observed that a lodestone would be attracted to any point on it, except on its equator.

Other points noted by Gilbert were that electric attraction was stopped by the interposition of a sheet of paper or the presence of moist air in the environment. The direction of the attraction was also always in a straight line towards the electric material, and that polishing the electric was not essential for demonstrating the effect.

Book 3: Of Directivity

In Book 3 attention is turned to the directive properties of the magnet. While coition, i.e. the coming together of the north and south poles of two magnets is easily observed, no similar coition is observed to take place between a magnet and the earth. The reason for this is that the strength of the earth's field is too small to demonstrate it. But if a small compass needle is placed at any point in the earth's field it will align itself along the direction of the earth's field at that point. How quickly it will do this will give a measure of what Gilbert calls the verticity of the magnet.

Applying this to Robert Norman's experiment where he suspended a small iron wire in a beaker of water, the wire showed both a vertical and a horizontal deflection. The vertical deflection Gilbert would also have observed from his experiments with a short wire on the terrella, and the angle to the vertical made by these deflections would have depended on the latitude of the wire placed on the terrella. Thus the vertical deflection of the versorium at the poles would have reduced to zero at the equator. From this, Gilbert deduced that magnets respond to the integrated effect of the earth as a whole. Norman's Respective Point was wrong, as was the theory of an extra-terrestrial source for all magnets. This was a major discovery of Gilbert.

Book 4: Of Variation

Book 4 is devoted almost entirely to the subject of variation, i.e. the difference in angle between the magnetic meridian and the geographic meridian. From Gilbert's acquaintance with the measurements made by ocean-going navigators, he knew there were irregular changes in this value, even over the Atlantic Ocean along a line of constant latitude. Similarly across the Pacific Ocean and other seas this variation existed, and it was known to increase as the poles were approached. Nevertheless there were known to be regions at sea where the variation was approximately zero. Moreover on each side of this zero the algebraic sign reversed. This result had initially given hope of being able to measure longitude as well as latitude from compass readings. But this hope turned out to be illusory, because of the irregularity of the results obtained in practice, using this technique. Nevertheless Gilbert believed that help with the Longitude problem could be obtained by recording experimental values of the variation in tabulated form, on a long journey along a given path. This was a method translated in the same year by Edward Wright. This emphasis on experimental measurements stands out as the dominant factor in De Magnete. Because of this its publication represents a turning point in the history of scientific literature.

Book 5: Of the Dip of the Magnetic Needle

It will be recalled that Robert Norman in 1581 discovered the existence of a dip in a magnetic needle. It pointed at an angle of 72 degrees to the horizontal in London. Gilbert generalised this result by showing how this angle varied as a function of latitude. He did this, not by travelling himself to different latitudes; instead, by making use of his terrella to represent the earth and a small compass needle, of the length of a barley-corn, in place of the longer needle used by Norman, a nomogram was devised to relate latitude to dip. This gave hope that by measuring the angle of dip on a sea voyage the latitude could be found. Although this was true for the terrella, the practicalities of the real earth with alternate sections of land and sea, made its use inaccurate in practice. But then Gilbert went further and asserted that every magnetic body imitated a soul "which provided the power for rotational movement". In the case of the earth this implied that its diurnal motion was due to its magnetism, rather than gravity as was discovered later.

Book 6: Of the Globe of the Earth as a Lodestone

Book 6 deals largely with rotation. Two of the bodies considered under rotation were the earth and the eighth sphere in which it was believed at the time all the stars rotated round the earth. In this view the earth was taken to be stationary. Although Gilbert himself does not quantify speeds of rotation in the two cases, this deficiency was made up by Edward Wright in his introduction to the book. He considers the speed of rotation of a point on the earth's equator as being only 0.25 miles per second if it is the earth that rotates. In contrast to this if it is the eighth sphere that rotates, with the earth remaining stationary, then a point on the equator of this sphere would have to travel at a speed of 5000 miles per second. In Gilbert's mind this was fairly compelling evidence that the earth rotated, rather than the eighth sphere with its fixed stars, and he thus became a cautious supporter of the Copernican view of cosmology.

Gilbert, however, denies Peter Peregrinus's view that a terrella poised on its poles in the meridian, will move circularly giving a complete rotation in 24 hours. He believed that the earth rotates in its diurnal motion because "if it did not the sun would hang with its constant height over a given point, and by long tarrying there would scorch the earth… and the uppermost surface of the earth would receive grievous hurt." And he goes on: "In other parts all would be horror, and all things frozen stiff with intense cold." "And as the earth herself cannot endure so pitiable and so horrid a state of things, with her astral magnetic mind she moves in a circle." "So the earth seeks and seeks the sun again, turns from him, follows him, by her wondrous magnetical energy." This appears to recall the idea of the soul of the earth providing the power for rotational motion as outlined in Book 5. It will, of course, be recalled that this was written before the time of Newton, who derived the correct cause of rotation in terms of gravitational rather than magnetic forces. But before the time of Newton there were other physicists who objected to the idea of the earth possessing a soul, among whom Renee Descartes must now be considered.

Rene Descartes (1596-1650)

Rene Descartes was the son of a French magistrate, and the grandson of a physician. His mother died in the year after he was born, and he was sent to the Jesuit School at La Flechc at age ten. This boarding school had a high academic reputation, particularly in mathematics, but this was combined with a caring concern for its pupils. As an example of this, Renee Descartes who was not considered to be physically strong, was allowed to stay in bed until late in the morning, after which he would then join the other pupils in class. It has been suggested that it was while lying in bed, gazing at the ceiling above him, that he conceived the idea of the three orthogonal coordinate axes, which were later to form the basis of Cartesian geometry.

After eight years at La Fleche he transferred to the University of Poitiers where he graduated in law in 1616. After a further two years he joined the army at the age of 22 as a gentleman volunteer in Holland, under Prince Maurice of Saxony. This did not, however, involve him in continuous military operations.

Rene Descartes's Vision

While in Germany on 10th November 1619 Rene Descartes had a vision which clarified for him what his objective in life should be. His life should be devoted to physics based on geometry, together with an interconnecting of all the other sciences. To enable him financially to carry this out, he sold his estate. Then, to give thanks for his life-determining vision, he made a pilgrimage to Our Lady of Loreto in Acona, Italy, where it is said that the house of the Holy Family at Nazareth had been miraculously transported by angels.

Descartes's Experiences in Europe (1621-1633)

Descartes then spent the next two years carrying out meteorological operations in Switzerland and Italy, followed by a further three years in Paris, making contact with scientists there. At age 32 in 1628 he moved to Holland, which had the reputation of allowing free thinking among its citizens. In that country he effectively spent the next nineteen years. After the first five years there, he had his first major work *Le Monde* ready for publication. It appears, however, that Descartes, like other scientists of the age, did not wish to be involved in the controversy of rival claims, and so delayed publication. But there were more delays to come. *Le Monde* dealt with both celestial and terrestrial physics, looked at from the Copernican point of view. But Catholic Cardinals had condemned in 1616 theories of a moving earth. And in 1633, the same year that *Le Monde* was ready for publication, Galileo in Rome, on his knees, had to make the following confession:

Galileo's Confession

"I, Galileo Galilei, son of the late Vincenzio Galilei of Florence, aged 70 years, being brought personally to judgment, and kneeling before you, Most Eminent and Most Reverent Lords Cardinals, General Inquisitors of the Universal Christian Commonwealth against heretical depravity, having before my eyes the Holy Gospels which I touch with my own hands, swear that I have always believed, and, with the help of God, will in future always believe every article which the Holy Catholic Church of Rome holds, teaches and preaches. Having held and believed that the Sun is the centre of the Universe and immovable, and that the earth is not the centre of the same and that it does move—I abjure that I will never more in future say, or assert anything, verbally or in writing which may give rise to a similar suspicion of me; but that if I shall know any heretic, or anyone suspected of heresy, I will denounce him to this Holy Office, or to the Inquisitor and Ordinary of the place in which I may be. I swear, moreover, and promise that I will fulfil and observe fully all the penances which have been or shall be laid on me by this Holy Office. But if it shall happen that I shall violate any of my said promises, oaths and protestations (which God avert!) subject myself to all the pains and punishments which have been decreed and promulgated by the sacred canons and other general and particular constitutions against delinquents of this description. So, may God help me, and His Holy Gospels, which I touch with my own hands. I, the above named Galileo Galilei have abjured, sworn, promised and bound myself as above; and in witness thereof, with my own hand, have subscribed this present writing of my abjuration, which I have recited word for word."

Despite this confession Galileo was confined to his house for the remainder of his life of nine years. Part of the original punishment was that he should also recite periodically the seven Penitential psalms, but he was apparently later absolved from having to do this.

Descartes's Publications

After Descartes stopped the publication of *Le Monde* in 1633 it remained unpublished until 1664, two years after his death. But other publications did follow. These included *Discourse on the Method of properly Guiding the Reason in the Search for Truth in the Sciences* (1637), and the *Meditations on First Philosophy, in which the Existence of God and the Distinction between Mind and Body are Demonstrated* (1642).

In these *Meditations* Descartes gives his two proofs of the existence of God as follows:

(1) The idea of God in a human mind must have been given by the Supreme Being, for a finite mind cannot conceive of an infinite thing.
(2) A perfect being must have the property of existence to be perfect.

He also lists the following Meditations:

(a) We may doubt the following things: Knowledge gained through the senses and dreams.
(b) The body is a mechanism of many parts. In contrast the soul and God are simple and indivisible.
(c) Man's limitations are such that although he can conceive of a geometric shape such as a chiliagon with one thousand sides, he cannot imagine it.

The French Government granted him a pension of 3000 livres towards the end of his life and the Queen of Sweden requested him to come and instruct her. After persuasion from the French Ambassador he accepted this invitation, but found that the Queen had fixed five o'clock in the morning for her lessons. Descartes caught a chill within a few months and died from pneumonia two weeks later.

In 1663 the Catholic Church banned his books, and even in his lifetime in Holland he met opposition from the Rector of the University of Utrecht. There Descartes was charged with atheism, and faced the possibility of seeing his books burned. But this was averted through the offices of the French Ambassador.

Descarte's Contribution to Magnetism

With regard to Descarte's contributions to magnetism, he refused to accept Gilbert's view that because, when a needle is magnetised, nothing physical appeared to be added to it, the magnet is animate and therefore behaves as though it had a soul. Descartes instead looked for a mechanical explanation of magnetism and produced the following argument. He suggested there were threaded grooves in lodestone and the earth. At one end these grooves were threaded so that a right-handed screw could travel along them, and at the other end a left-handed screw which would block a right-handed thread. He then suggested that grooved particles from the sun would enter with a screw direction at the North Pole opposite to that at the South Pole because of the uni-directional rotation of the earth. These grooved particles entering at one pole travel through the earth to the other pole. When they emerge they return to their original pole through the air which has resisted and deflected them. Descartes described this motion as a kind of vortex, believing that it explained the magnetic fields round the earth and other magnets, as forming closed loops.

Although Descartes got this wrong, it was a mechanical picture which appealed as something that could be physically understood. He believed the whole of space must be filled with particles of some kind which could transit forces between them, so that he denied action at a distance. In this way he thought that all occurrences in the natural world could be predicted by calculations. Thus he did not realise that the truths of physics have to be built up experimentally, before theories can be devised to embrace them. Among the people who soon contributed to this development in an experimental way was the Irishman Robert Boyle.

Robert Boyle (1626-1691)

Robert Boyle was the fourteenth child of the Earl of Cork, who had been given land in Ireland by Queen Elizabeth as part of her political policy of subduing Irish Chiefs and distributing their land among Englishmen. Although his father had landed in Ireland in 1588 with only £27 in cash, a diamond ring and a gold ring given to him by his mother, he eventually became the greatest landowner in Ireland, including some land there which he bought from Sir Francis Drake. He ran a patriarchal household, arranging marriages for his children, and made friends with royalty to the extent that King Charles I gave away a bride to Robert's next eldest brother, who was married at age fifteen. When Robert was aged fourteen his father presented a gold and diamond ring to a girl whom he considered to be a suitable match for Robert, but this pledge was not fulfilled and Robert in fact never married.

When Robert was three years old he was entrusted, as the custom was, to a country nurse, and he never saw his mother again. She died when he was still three, and he regretted never having known her. But the family custom was to remember each year the day of her death, as a day of solemn mourning in the family home.

At age five Robert returned home and was educated by a French tutor who also taught Latin. After four years at home he was sent in 1635 to Eton College, where the Provost was Sir Henry Wooton. Sir Henry had earlier been the English Ambassador in Venice, and was familiar with the works of Paolo Sarpi, the author of *The History of the Council of Trent*. But perhaps Sir Henry is better known for his incautious definition of an ambassador, as "An ambassador is an honest man sent to lie abroad for the good of his country."

Most of Robert's education was in the hands of a Fellow of Eton, who early recognised Robert's ability, and taught him individually. But Robert loathed school games. After four years at Eton The Earl of Cork withdrew both Robert and his elder brother Francis from the College. He sent then, under the care of a Frenchman, Isaac Marcombes, a French Calvinist settled in Geneva, to visit Rouen, Paris and Geneva.

Robert Boyle's Religious Experience in Geneva

It was during the visit to Geneva, surrounded by the Alps, that Robert had a terrifying religious experience during a night of thunder and lightning. As a result he made a vow to dedicate himself to piety. He had been made aware of his personal insufficiency. For some time thereafter he was unable to take Communion, and even contemplated suicide, but eventually a sense of reconciliation came and he was able to begin his religious life again, but on a new foundation.

Robert Boyle's Travels Thereafter

The following year at age fifteen he visited Italy. Before he left England his father had obtained from King Charles a letter requesting all foreign authorities to provide all such assistance as was necessary, and a separate letter authorising him to visit Rome. In Italy he read about Galileo. While they were in Florence Galileo died nearby, and a tomb was erected for him at the public expense. In Rome, Robert said he was French to avoid trouble from English priests, since the Earl of Cork's family were staunchly Protestant. Nevertheless in Rome Robert was the victim of a homosexual attack from two Friars, from which he was able to extricate himself.

It was on their way back from Italy in 1641 that unexpected difficulties arose for Marcombes, Robert and his brother Francis. From the time that they had left England the Earl of Cork had provided a sum of £1000 p.a. to finance their travels. But in 1641 a rebellion had broken out in Ireland which meant that the Earl had to purchase armaments to protect his property, and there was insufficient money available to cover the costs of travel in Europe. This meant that the French Tutor Marcombes, had to provide support for a number of years for the two boys. The older brother succeeded in reaching England first, but Robert did not return until 1644 at age eighteen. There he was able to stay with his favourite sister Katherine, who had contacts with the Parliamentary forces, who were eventually able to regain control after the Battle of Naseby the following year. It appears that as a result of her influence Robert was able to obtain favourable treatment of his estates in both England and Ireland.

The estate in England which Robert inherited from his father was at Stalbridge in Dorset, and it was here that Robert began his chemical experiments. Initially these would necessarily have involved repeating other people's experiments. But the hopeful eventual aim was to achieve the transmutation of metals. Robert Boyle's view of the possibility of changing metals into gold was analogous in his view to the redemption of mankind in theology. The care with which he carried out his experiments, both personally and through his assistants, led to his association with weekly meetings on natural and experimental philosophy in London. His experimental care in chemistry was also transferred to practising medical care on himself. He kept a number of cloaks, with the choice of the one he ought to use each day being decided by the reading from his thermometer. In present day language, he might be described as a hypochondriac, but this would be an unfair description for the age in which he lived.

Robert Boyle's Religious Romance: The Martyrdom of Theodora and Didymus

In addition to his chemical experiments, Robert at age twenty wrote what is possibly the first religious romance, *The Martyrdom of Theodorus and Didymus*. Dr Johnson is quoted as saying that on the basis of this work Boyle might have become a great writer. The book is based on the martyrology of Theodora of Antioch, which in the original classical version is said to have covered only about one page. But Robert Boyle extended this by making Theodora not only virtuous but attractive, and able to produce a logical argument against marriage. The reason for this was possibly to show that human love is not the greatest ideal.

In Boyle's version of the romance the beautiful Theodora is sentenced by the ruler of Antioch to choose between offering sacrifices to pagan gods or being prostituted in the public brothel. Choosing the latter, she is led away and before the first customer is admitted, resolves to resist so vigorously that he will murder her. In this way she will also avoid the sin of suicide.

Her lover Didymus decides to rescue her. He finds an officer Septimus who is willing to help and pays a large reward to him. As a result Didymus is admitted to her cell and tells her how she can escape in Didymus's clothes. But Theodora refuses and implores Didymus to kill her. He refuses and

Theodora agrees to put on Didymus's clothes, and escapes to a female friend Irene. Theodora tells Irene that if she were to marry she would have to put her whole heart into that relationship, to the detriment of her life of devotion to God alone.

News comes that the escape has been discovered. Didymus is sentenced to execution within one hour. Theodora appeals to the judge that she should suffer in Didymus's place. The judge orders them both to be killed. Didymus is executed and his body is shown to Theodora who reacts by being proud of his suffering. She makes a final speech from the scaffold. And her glad soul was carried by the angels to heaven.

Robert Boyle's Religious Views

For six years Robert Boyle stayed at Stalbridge. In addition to his chemical experiments he used his time to learn Hebrew, and it is said that he learned a Hebrew grammar almost by heart. And he learned Chaldee grammar to appreciate better the book of Daniel and other portions of the Old Testament. He added a Syriac grammar to read the divine discourses of our Saviour in his own language.

These accomplishments, added to his natural ability, combined to produce a number of sayings which are worthy of reproduction:

> "There are two chief ways to arrive at the knowledge of God's attributes; the contemplation of his Works and the study of His Word.

> Science is primarily a means of illuminating the imperfect knowledge of God which mankind has received through the Scriptures.

> Religious doubts are like toothache, not fatal but very troublesome.

> Lord, when I lose a friend, teach me to settle all my love on Thee.

> Every creature is a preacher, and in the study of nature the deity is revealed.

God holds out the offer of salvation, and His Word in the Bible encourages man to reach upwards. But it is far beyond his unaided attainment. Yet if we truly strive for it, God will stoop to give us the prize which we cannot reach without His help.

He alone loves God as much as he ought, that loving Him as much as he can, strives to repair the deplored imperfection of that love, with an extreme regret at finding his love no greater.

Human love is chancy, and is devotion misaddressed, for when you talk of offering up of hearts, though it is said to her, it is meant to a divinity

It does not make any difference to God's love if man rejects it.

The problem of the suffering of the innocent in connection with divine justice, is as old as Job.

The world of nature was the handiwork of the Creator. The physical world is God's other Bible."

Robert Boyle's Further Experimental Skills

After six years at Stalbridge, Robert Boyle visited Ireland in 1652 in order to agree his own share of his father's estate. He stayed there about one year, benefiting from the presence and skills of Dr Patty who had hosted meetings of the natural and experimental philosophers' group when it met in Oxford rather than in London. Dr Patty had then been appointed Chief Physician to the Commonwealth Forces in Ireland, and had skills both in surgery and in the design of boats. From Dr Patty it was possible for Robert Boyle to extend his chemical skills to those more useful in his future research interests in air, magnetism and electricity. From a financial point of view this became possible since his share from his father's estate brought him an income of £3000 p.a. With this he moved to Oxford in 1652.

In 1658 Gaspar Schott published a book describing von Guericke's air pump, and from it Robert Boyle, along with his inventive assistant Robert Hooke, devised a more effective pump. From this the elasticity of the air was demonstrated, together with the relation between the pressure and volume of a gas, known as Boyle's Law.

The Restoration of the Monarchy in England in 1660

The restoration of the Monarchy in England in 1660 under Charles II, brought the offer of a peerage to Robert Boyle, but he declined this. Likewise he was invited to take Holy Orders, with the implication that this would lead to a Bishoprick, but again this was declined, partly on the stated ground "that he had no inward notion to it by the Holy Ghost".

Three years afterwards on the twenty second of April 1663, the Royal Society was incorporated by King Charles, who displayed personally and through his court great interest in its proceedings. It even became customary for ladies of the Court to attend some meetings. The Royal Society, early in its proceedings, prepared a set of twenty demonstrations for a visit to be attended by King Charles, and of these twenty, nine were arranged by Robert Boyle, of which two were magnetic. These were:

(1) Destroying the attractive virtue of a lodestone by heating it red-hot, while its directive properties are maintained.
(2) Reversing the polarity of an iron rod by making one end red-hot and refrigerating the other end, and then destroying the magnetism by striking it in the middle.

Although Robert Boyle had declined social and ecclesiastical honours in the past he did accept a Doctor of Physics degree from Oxford in 1665 at the age of thirty-nine. In the following year, however, he declined the offer of being Provost of Eton, his old school. In 1669 he left Oxford and went to live in London with his sister Katherine, with whom he had much in common. They began each day with meditation and private devotions, and then Boyle went to his laboratory. For the rest of his life which was spent in London, Boyle continued with his work on both science and religion.

Robert Boyle's Religious Activities in London (1669-1691)

(1) He was in charge of the translation of the Four Gospels and the Acts of the Apostles into the Malay Language which were sent over all the East Indies.

(2) He gave a noble reward to the translator of Grotius' Dutch treatise of the Truth of the Christian into Arabic.

(3) He had a large share in the production of the New Testament into Turkish.

(4) He contributed largely to the impressions of the Welsh Bible.

(5) He contributed to the Irish Bible for its use in Ireland, and further for its use in the Highlands of Scotland. This involved sending two hundred copies there in 1687/88, which made one book for every parish in the Highlands. It was said that about one half of the ministers preached only in Irish. And as to women and children in the Highlands, scarce one in twenty could speak English. The zeal for the Scriptures in the Highlands was such that the people sent for the Bible sometimes to one part of the Parish, and sometimes to another, that they might read it on the weekdays; and then they returned it to the Church on the Lord's Day, that all might hear it read publicly.

(6) He gave £300 to advance the design of propagating the Christian religion in America.

One ministerial correspondent with Robert Boyle offered to pay for the binding of additional unbound copies, and to distribute them subject to two conditions:

(a) That the Minister read some chapters each Lord's Day to those who had never had a chapter read in their lives, and on other occasions, such as baptisms, burials, marriages etc.

(b) That if the Minister leaves the Parish, the Bible remains in the Parish.

Robert Boyle's Experiments on Magnetism in London

Robert Boyle would have been familiar with Gilbert's recorded experiments on magnetism. He would have repeated many of them, and still found fifteen which are recorded as in some way sufficiently different to merit inclusion in his collected works. Of these the following are noted:

(1) A piece of steel, fitly shaped and well excited, will like a lodestone have its determinate poles. It will draw other pieces of iron and steel to it, and communicate the same kind, though not the same kind of attractive, directive virtue, as settled and durable powers for many years, if the lodestone which magnetised it were vigorous enough. I have seen one that yielded an income to its owner, who received money from navigators for suffering them to touch their needles, swords, knives etc. In a piece of iron thus excited it retains its malleability as before, as an ordinary piece of the same metal unexcited. If this disposition of the excited iron be destroyed, though the form of the metal be unaltered, the former power of attraction will be abolished as when an excited iron is made red hot and cooled again.

(2) That a lodestone may be easily deprived by ignition of attracting material bodies, and yet be scarce visibly changed, but continue a true lodestone in other capacities, and the lodestone thus spoiled may have its poles altered at pleasure, like a piece of iron. But in a natural lodestone, never injured by fire, the attractive property, the directive property, and the poles may be changed by taking a very small fragment, and by applying a pole of a small but vigorous lodestone to it. In this way I could in a few minutes change the poles of the small fragment, but only for a small fragment.

(3) In shops of smiths and metal turners, when hardened tools are treated by attrition, if while thus warmed you apply them to filings or chips they will take them up as if they were touched by a lodestone, but they will not do so unless excited by rubbing till warmed. And the filings will continue to stick after they grow cold again.

(4) The iron bars of windows in an erect position may grow magnetic so that the North pole of an excited needle applied to the bottom of the bar will drive it away, but attract the South pole.

(5) Keeping a lodestone red hot, though you cool it in a perpendicular position you may deprive it of its magnetic attraction. But a rod of iron, heated and cooled in a perpendicular position, acquires a manifest verticity.

Robert Boyle's Experiments on Electricity in London

As with magnetism, Robert Boyle would have owed much to Gilbert for his pioneering work on electricity. But he made his own contributions as the following selection shows.

(1) My choicest piece of amber draws not only sand and mineral powders, but filings of steel and copper and beaten gold itself provided they be minute enough.

(2) After evaporating one quarter of good turpentine, the remaining body would not when cold continue a liquid, but hardened into transparent gum, almost like amber, which proved electrical.

(3) Mixing two liquids, petroleum and strong spirit of nitre in a certain proportion, and then distilling them till they formed a dry mass, this left a brittle substance as black as jet, and glossy, which proved electrical.

(4) When antimony was burned to ashes and made into a transparent glass, this proved electrical when rubbed.

(5) False locks of some hair, being by curling or otherwise brought to a degree of dryness will be attracted by the flesh of some people.

(6) A large and vigorous piece of amber, and a downy feather from the body (not the wings or tails) of a large chicken were selected. When the amber was excited the neighbouring part of the feather was drawn to and stuck fast to it, but the remoter parts did not move. Boyle's forefinger was then offered to the erected downy feather, which attached itself to his forefinger, then to a cylinder of silver likewise, and an iron key, with the same result.

Boyle died in 1691 and was buried in Westminster Abbey. He left to the Royal Society his collection of ores and specimens, with these words:

"Wishing them also a happy success in their laudable attempts to discover the true nature of the works of God, and praying that they, and all other searchers after physical truth, may cordially refer their attainments to the glory of the great Author of Nature, and to the comfort of mankind."

In his will drawn up in 1691 he left a sum of fifty pounds a year to fund an annual set of eight lectures for proving the Christian religion against notorious Infidels, viz. Atheists, Theists, Pagans, Jews and Mahometans, not descending to any controversies that are among Christians themselves. The first series of lectures was given by Mr Richard Bentley, later Master of

Trinity College, Cambridge, with the Title "A Confutation of Atheism". In the last of this first series his concluding argument was the demonstration of the necessity of a Divine Providence from the constitution of the universe as demonstrated in Newton's *Principia*, which had been published four years before. These annual lectures were continued from 1692 to 1732, and have recently been revived in 2004.The reference to the *Principia* brings in the name of Edmund Halley, to whom credit must be given for its very existence, and whose contribution to magnetism will next be considered.

Edmund Halley (1656-1742)

Like Robert Boyle, Edmund Halley was born into a wealthy family. His father was a property owner, and Edmund was educated at St Paul's School in London. This was followed by study at Queen's College in Cambridge. His name is chiefly remembered now in association with Halley's Comet, which he observed in 1682, but which was actually observed two weeks before in Maryland. His calculation of the comet's return in 1759, after his own death, was a triumph of the validity of Newton's theory of gravitation.

His mathematical ability was such that at age twenty he had published three papers in the Philosophical Transactions of the Royal Society. But at age nineteen he introduced himself to the first Astronomer Royal, John Flamsteed, who had been appointed by Charles II to form the Greenwich Observatory established in 1675. From his salary of £100 a year Flamsteed had to supply most of his own instruments, with which he formed the first catalogue of fixed stars. It was therefore perhaps not too surprising that he accepted Halley's letter advertising his own experimental measurement capability. Flamsteed's plan was to map the Northern stars, and Halley decided to do the same for the Southern stars. To do this required a station in the southern hemisphere, and he obtained an introduction to the East India Company through Charles II. He left Oxford without a degree in 1676 and sailed to St Helena rather than Brazil, because English was spoken there. His catalogue contained three hundred stars in the southern hemisphere.

In 1678 Halley returned to England, and a friendship with Sir Isaac Newton developed which resulted in the publication of Newton's *Principia* at Halley's expense. It has been said that but for Halley "in all probability that work would not have been thought of, nor when thought of written, nor when written printed". In 1684 Hooke, Halley and Christopher Wren realised that Kepler's Laws of the planets moving round the sun could be explained mathematically by the inverse square law. The problem was to show this analytically. On Halley's next visit he made Newton promise to write a book on it. In eighteen months he did. It was the *Principia*.

In 1691 Halley's father got into financial difficulties, so Halley applied for the Chair of Astronomy at Oxford. This was unsuccessful, possibly because he was thought to be a religious heretic. Accordingly in his approach to magnetism he decided to make his magnetic model of the earth such that it did not oppose the Christian viewpoint of the universe not being eternal. Earlier he had thought, as Gilbert had done, of the whole earth being one large magnet, but in Halley's mind this magnet could have four poles. Two of these were in the northern hemisphere, and two in the southern. The poles were of different strengths and the experimental result of zero magnetic variation across a diagonal in the Atlantic could be explained by one pole being in northern Europe and the other in North America. This was Halley's first magnetic model. But eight years later he produced another model in1691. He now proposed that the earth had an outer crust five hundred miles thick, with three concentric magnetic globes with diameters proportional to Venus, Mars and Mercury. Inside these was a solid ball two thousand miles in diameter, with the spaces between the shells being inhabited. It was not specified what kind of inhabitants these might be, but at that time Christians, including Robert Boyle, are believed not to have ruled out the possible existence of fairyland. Correspondence exists between Robert Kirk, the Episcopalian Minister of Aberfoyle, who wrote a book *The Secret Commonwealth of Elves, Fauns and Fairies*, and Robert Boyle.

To provide a finite ending to the earth Halley postulated that a cloud of gas might be gradually slowing down all the heavenly bodies. This at least made the age of the earth finite. In 1693 Halley published also Tables of Mortality, one of the first to provide annuities based on fact. In 1701 he published the first Atlantic Magnetic Chart, constructed from fifty-five available observations. In 1703 he was appointed Professor of Geometry at Oxford, and when Flamsteed died in 1720, Halley was appointed Astronomer Royal.

Stephen Gray (died 1736)

All previous contributors to the development of magnetism and electricity have benefited from connections with either financial wealth, or links with the

Church which have provided them with at least time in which to carry out experiments. In most cases also the dates of their birth and death have been known at least approximately. But Stephen Gray's origin and birth date appear not to be known. What is known is that he resided late in life at Charterhouse, a charitable foundation in London, built originally for Carthusian monks. This was later converted into a hospital for eighty pensioners and a school for forty boys. The elderly residents were required to be over fifty years of age and members of the Church of England. In the nineteen fifties the number of pensioners had fallen to forty, but the residential school today has over seven hundred pupils. Previous pupils of the school have included Joseph Addison, John Wesley and Lord Baden Powell.

Stephen Gray is known today only for his publications in the Philosophical Transactions of the Royal Society, to which he contributed research papers between 1720 and 1736. Although he lived in a charitable foundation there is record of his having two friends in better financial circumstances, one of whom was a Fellow of the Royal Society in London. It appears that his experimental work started by following the work of Robert Boyle who investigated among other things the electric attraction between the human cheek and a woman's long hair. Stephen Gray said that he made leather and parchment and paper and hair and feathers and threads all electrical by rubbing them. He then asked himself if their electrical effect could be transferred to other objects connected physically to the original electric source. To answer this he took a long glass tube and put corks in the ends. He rubbed the tube and found that feathers were attracted to both the glass tube and the corks at its ends. Into the cork he then inserts a wooden rod of increasing length, terminated by a variety of objects, and the same result is obtained, even when the rod is replaced by a long fishing rod. When the rod was replaced by a horizontal packthread supported by vertical silk threads, a transmission distance of over six hundred feet was obtained. He then suspended a boy and found that when the tube was rubbed, and held to the boy's feet the effect could be felt on the boy's face, thereby indicating the conductivity of the human body.

Another of Gray's results of importance was his discovery from the use of solid and hollow wooden cubes that the charge resides on the surface of the electrified body, for no part but the surface attracts. It appears that some of Gray's results were communicated to the Royal Society by J. T. Desagulier (1683-1744), the son of a Huguenot pastor who escaped from France after the revocation of the Edict of Nantes by Louis XIV in1685, and who was afterwards ordained. Desagulier also translated into English the work of W.J.'s Gravesande (1688-1742), the Dutch Scientist who wrote in 1720 a very clear work on Natural Philosophy with very full experimental details.

Charles-Francais du Fay (1698-1739)

Du Fay was Superintendent of Gardens to the King of France. He was reputed to be of a very equable temper, and was on good terms with Stephen Gray who has been described as irascible. Du Fay found that when gold leaf, electrified by excited glass was brought close to an excited piece of resin that an attraction occurred between them. He had expected the opposite and thus discovered the existence of two kinds of electricity, vitreous and resinous. Each of them repels electricity of the same nature as their own, but attracts electricity of the opposite kind. When Du Fay sent a copy of his researches to England for presentation to the Royal Society, he also included a copy for Stephen Gray "who works in this subject with such application and success. And to whom I acknowledge myself indebted for the discoveries I have made, as well as for those I may possibly make hereafter." This was a generous tribute from the French scientist who had contributed to French Academy investigations in every one of the six subject areas recognised by them, viz. chemistry, anatomy, botany, geometry, astronomy and physics. DuFay also repeated Gray's experiment of suspending a boy and discovering his conductivity. He then extended this by suspending himself and described the sensation caused by an electric tube held close to his face as similar to having spider's web drawn over it, and feels the pricks and burns of the electric sparks as they dart from his fingers. Gray in turn conducted a similar experiment and commented that the sparks seemed to be of the same nature as those connected with thunder and lightning. When Gray died in1736 Du Fay gave a final tribute to him: "He was almost alone in England in pursuing his subject. To him we owe the most remarkable discoveries pertaining to it, so all those who love nature and her work must infinitely regret him."

The next chronological contributions to be considered will be those from Germany and the New World.

CHAPTER THREE

EULER'S CLARITY OF MIND

(c.1740 A.D. – c .1800 A.D.) (SECTION 1)

WITHIN THE TIME SCALE of Chapter Two, which broadly covered developments for both Electricity and Magnetism up to 1740, there were a number of ideas and associated experiments which came to fruition in the following sixty years. Among the first of the scientists responsible for some of these developments was the German Otto Von Guericke.

Otto Von Guericke (1602-1686)

Von Guericke was born at Magdeburg in Prussian Saxony. His father and grandfather had both become mayors of this Lutheran town, and he himself was privileged to have studied at the Universities of Leipzig, and Jena, before going to Leiden to become proficient in Engineering and Modern Languages. His competence in Mathematics was such that in later life as an astronomer he was the first person to predict the periodicity of the return of comets. After these studies he returned to Magdeburg in 1626 to marry and to become an alderman in that city.

Von Guericke and the Thirty Years War

It will be recalled that the Thirty Years War started in 1618, following earlier rival groupings of Calvinists and Catholics. The trigger which began the war was the succession of Ferdinand of Styria to the Hapsburg dominions. He has been described as a bigoted Catholic, and the Protestants in Bohemia rebelled, throwing, it is said, "two Regents of the kingdom out of a window in the palace at Prague". From this Bohemian revolution there developed a European war which eventually involved James VI and I, Charles I, Christian IV of Denmark, Sweden and France. The intervention of Sweden by Gustavus Adolphus in 1630 to support the Protestant side was successful at Breitenfeld (1631), but this came too late to prevent the sacking of Magdeburg. It is said that 25,000 people, five sixth of the population, perished when the Catholic troops overran the barricades, or when fire consumed the wooden town. Von Guericke was able to buy his own freedom, went initially to Sweden, but returned to Magdeburg the following year. Sixteen years later in 1648 the war ended with the Peace of Westphalia, at which Von Guericke represented Magdeburg. One of the results of this peace was that German Calvinists obtained legal recognition of their theological viewpoint. After the war Von Guerick was able to devote himself to scientific pursuits. Initially these were in the field of pneumatics and by 1650 he had invented the air-pump, as a means of producing a controlled vacuum.

Von Guericke and Electricity

Von Guerick was, of course, familiar with Gilbert's *De Magnete* and his association of electricity with gravity. But he did not accept Gilbert's view that the earth was a magnet. Instead he appears to have believed that sulphur was a substantive part of the earth's composition, along with several other minerals. What these minerals were and their various proportions in the total composition is not clear. Von Guerick himself is said to have distributed globes and given lessons in their manufacture. But although he himself produced (in 1660) a sulphur globe about the size of a child's head which when rubbed attracted small particles of matter, this demonstration does not appear to have been taken up by others. But one feature of it was significant. If the particle was a feather which touched the sulphur globe, then some electric charge on the globe transferred to the feather and produced repulsion on it. In this situation moving the globe also moved the feather, making an attractive demonstration. Further, the same face of the feather is kept unchanged if rotation of the globe is applied, just as the moon always turns the same face towards the earth. In the case of heavier particles this effect would be less likely to occur because of their heavier mass, so it was easier to demonstrate the attractive property with a feather. Moreover, the globe would then retain them just as the earth does in its daily rotation.

In the manufacture of the sulphur globe Von Guericke took a glass phial and filled it with finely ground sulphur. This was then heated, and then cooled, and the glass was broken leaving the sulphur globe. What was later discovered by Isaac Newton was that the glass sphere on its own was sufficient to produce the electricity without the sulphur. The consequences and developments of this discovery were worked out by the Englishman Francis Hauksbee who appears to have been an assistant to Isaac Newton at one time.

Francis Hauksbee (1666-1713)

Little seems to be known about the early life of Francis Hauksbee. He had an instrument maker's business in London, selling air pumps and barometers, and there is no doubt that he was a hands-on experimentalist. Soon after Isaac Newton was elected President of the Royal Society in 1703 he arranged that Francis Hauksbee should be appointed Curator of Experiments. This involved him demonstrating experiments to the Fellows. About the same time he also began to give public demonstrations at his shop to those who were willing to pay for these. An unusual characteristic of these public demonstrations was that he obtained lecturers to do the talking while he did the demonstrating. The lecturers included the Master of the Mathematical School at Christ's Hospital, London, and even more significantly the Lucasian Professor of Mathematics at Cambridge University, William Whiston.

William Whiston had succeeded Isaac Newton in the Lucasian chair at Cambridge in 1704, and was the Boyle Lecturer in 1707. Consequently he was presumably a respected Anglican Christian at that time. His theological studies of the Apostolic Constitution thereafter led him to the view that Arianism was the Creed of the Primitive Church. This belief was based on the teaching of Arius, a Presbyter of Alexandria, in 318, that Christ was created and made. But at the Council of Nicea in 324 the Church laid down that "Christ was the Son of God, the only Begotten of the Father, of the Substance of the Father and very God of very God". The Christian Church therefore opposed Whiston's views and he was expelled as a heretic by Cambridge University in 1710. Thereafter Whiston wrote a five-volume work entitled *Primitive Christianity Revived*, and produced a reformed liturgy. His theological studies were then devoted to translating from the Hebrew the historical works of Flavius Josephus, the author of *The Antiquities of the Jews* and *The Wars of the Jews*.

Hauksbee's Work in Electricity

Hauksbee's original demonstrations involved rubbing an evacuated glass globe of about 9 inches diameter, and then allowing air to return to the globe. But for the purpose of our survey the rubbing of an air filled flask or globe with either one hand or a leather cushion is sufficient for us to consider. The resultant effect is the production of an electric charge on the surface of the globe. This is manifested by the presence of sparks on the surface. This charge was collected with a metal chain resting on the globe.

Historically these sparks made a lively demonstration through being transferred to a boy who was suspended horizontally with his toes about an inch from the globe, and seeing them appear from his fingers. It was the normal practice to have the globe rotated at high speed, and the light produced in this way could make letters visible in a darkened room.

Because large diameter globes would not easily be available, it was natural to extend the experiment to examine the effect of rubbing glass rods and glass tubes instead of globes. Again luminescence, sparks, attraction and repulsion of light brass leaves were observed.

Hauksbee's first paper was published by the Royal Society in 1706. Over the following years he had a number of other papers published by them, and eventually he was made a Fellow, but died at the early age of 47 in 1713. After his death some of his public lectures were carried on by Desagulier, who also received the appointment of Curator of Experiments at the Royal Society.

In Holland Hauksbee's machine was described in a book published around 1720 by 's Gravisande (1688-1742), Professor of Astronomy and Mathematics at Leyden University. This book was translated into English by Desagulier. The chair at Leyden University was to become famous in 1746 with the invention of the Leyden Jar associated with the name of Professor

Musschenbroek, who succeeded 's Gravisande there. But there were earlier advances to be made in Germany, and these will be considered next.

Professor Georg Matthias Bose (1710-1761)

Bose was born in Leipzig, the son of a judge. He was intellectually able, hardworking and appears to have been a natural publicist. He began his academic career as an assistant lecturer in mathematics and physics at Leipzig University, founded in 1409 by a secession of students from the University of Prague. In Leipzig opposite the entrance to the Johanneskirche stands the Reformation Monument with statues of Luther and Melanchthon. Whether this had any influence on Bose is not known, but it did not prevent him later from making known a connection with Pope Benedict XIV after he was appointed as Professor at the University of Wittenberg. This connection apparently displeased the Faculty of Theology in Wittenberg, the birthplace of the Reformation.

Bose's Work in Electricity

Bose's experimental interest in electricity was based ultimately on the work of Hawsbee, from whom he borrowed the idea of using a spherical globe, rather than a cylindrical tube. This provided him with much more electrification than he anticipated, to the extent that he communicated the idea to the Royal Society in London, through William Watson, who was possibly the leading electrical physicist in England at the time. For the collection of the charge produced by the global generator Bose used initially either a suspended person with his feet almost touching the rotating globe, or a person standing on wax holding his hand close to the globe. Later and more simply, the suspended person was replaced by a metal tube twenty-one feet long and four inches in diameter. To prevent damage to the globe from contact with the metal tube a bundle of threads was placed between these two components.

The purpose of the metal tube was to produce an increase of charge similar to that produced by a larger diameter globe. Increasing the length of the tube increases the capacitance. Similarly increasing the diameter does the same though to a lesser extent. The charge distribution along the tube will not be uniform but will increase at both ends. This is the standard edge effect which exists also with current electricity, which occurs when a change is made from an infinite to a finite length of conductor. A metal rod can alternatively be used in place of the tube.

In 1744 the German Academy of Sciences received an address from Dr Christian Friedrich Ludolff, at which he demonstrated sparks from the far end of a metal rod entering a container of water. When this water was replaced with an alcohol it burst into flames, much to the amazement of the audience. The combustion of the alcohol then naturally opened the way to combustion of other materials. To complete the story it has to be said that Professor Bose thereafter made a claim to prior discovery. And this has been a common feature in a number of discoveries in electricity and elsewhere. Thus Andrew Gordon, a Scottish Benedictine monk in Erfurt used a glass cylinder in place of the globe used by Hawksbee and Bose. He also invented an electric bell by making use of two gongs which were struck alternately by a metal sphere suspended by a silk thread. When this was electrified it struck one gong, from which it was repelled to strike the other. He also used electricity to kill small birds, as did a number of other experimenters.

The Uses of Electricity

With regard to what use electricity might have for people in general, it was remarked by Johann Gottlob Kruger, Professor of Philosophy and Medicine at Halle, that the philosopher's life consisted of "trying to understand what you do not see (like electricity), and not believing what you do." With regard to what use electricity may be to anyone, he said, "No use has been found in Philosophy or Jurisprudence, and therefore where else can the use be but in medicine?" He proceeded to urge his students to investigate, and 1744 in the same University Christian Gottlieb Kratzenstein made the first experiments on the living human body over a wide range of functions.

This newfound interest in electricity in Germany came to the knowledge of natural philosophers in countries which up till then had been in the forefront of such knowledge themselves. In particular the Royal Society in London was kept informed by William Watson, whose contributions will now be discussed.

William Watson (1715-1787)

William Watson was born near Smithfield, in London, the son of a reputable retailer of grain. When his father died young he was sent for his education to the Merchant Taylors' School in London.

Merchant Taylors' School

This was founded in 1540 as a grammar school which was to educate 250 boys. Of these, 100 were to be poor, provided "that they were apt to teach", and they would pay no fees. In addition another 50 were to be able pupils, also poor men's children who would pay two shillings and two pence per quarter. The remaining 100 would pay five shillings per quarter.

The scholars were examined in Latin, Greek and Hebrew, and were required to learn the Catechism in English or Latin. No cock fighting or tennis play was allowed. The High Master was to direct in doctrine, learning and teaching. He should be a man "in body whole, sober, discrete, honest, virtuous, and learned in Latin and Greek if such can be gotten". When the first High Master was appointed the scholars poured in. The school was later to rank itself fourth in Britain, after Westminster, Eton and Winchester. As some indication of the accuracy of this claim, five out of the fifty-four translators of the Bible (1604-1611) were Merchant Taylors' men.

William Watson as Apothecary and Scientist

After leaving Merchant Taylors' School, William Watson was apprenticed to an Apothecary in 1730, aged fifteen. The apprenticeship was to last eight years. At this time the Apothecaries were benefiting from an enactment of 1704 which granted them the right to prescribe as well as to make up medicines. But even for the compounding of medicines it was clearly necessary for them to be able to identify plants in situ and to learn the chemistry associated with the mixing of the component herbs. It is recorded that he made frequent early morning excursions from London to find such plants and won a prize from the Apothecaries Society given annually to young men exhibiting superiority in the knowledge of plants.

In 1738 he married and set up his own home. Through his botanical contacts he attended, in the same year, meetings of the Royal Society, and in 1741 was elected a Fellow of that Society. Throughout his life his published contributions to its Proceedings naturally included botanical subjects, such as the death of two French prisoners from eating the roots of Hemlock Dropwort from which they suffered initially from lockjaw. He also published papers on the sexes of plants, on the poisonous effect of the yew tree on horses, and on the animal origin of coral, which was previously regarded as a vegetable.

William Watson and Electricity

Watson's most important work centred on electricity. In 1744 he was the first person in England to set fire to a spirit of wine, which he then extended to all the inflammable liquids he had in stock. This contributed to his being awarded the Copley Medal, the highest Royal Society award in 1745. The following year he conceived the idea that if he insulated himself from the ground he might be able to generate more electricity by rubbing the glass tube. Because this gave a negative result it led to the conclusion that the electricity came from the ground, with the rubbing of the glass performing only as a pump action and not as a generator.

At a personal level Watson was a very clubbable man. Possibly as a result of this he contributed to bringing to the notice of natural philosophers in Europe the electrical discoveries of Benjamin Franklin in America. In particular he reviewed Franklin's first communication before the Royal Society in 1748. In 1747/48 he was the principal actor in the famous experiments, in one of which the electrical circuit was extended to four miles to find the velocity of electricity. From their measurements they decided this was infinite. In 1759 he became a licentiate of the Royal College of Physicians, having discharged his membership of the Society of Apothecaries, after paying a fee of £50 for this discharge. As Physician to the Foundling Hospital he wrote a pamphlet comparing methods of inoculating against smallpox, from which Benjamin Franklin's son died in 1736. He also became Vice-President of the Royal Society, and received a knighthood in in1786. The contributions of Benjamin Franklin and his School to the understanding of electricity will be considered next.

Benjamin Franklin (1706-1790) and his School of Electricitty, Comprising, F.U.T. Aepinus (1724-1802), John Robison (1739-1805 and Joseph Priestley (1733-1804)

Benjamin Franklin

It was said of Franklin by the French economist and statesman A. R.J. Turgot, "He snatched the lightning shaft from heaven, and the sceptre from tyrants." These vignettes refer to two only of Franklin's accomplishments. He has also been described as a printer, author, publisher, inventor, scientist, public servant and diplomat, which also represents the chronological order of his career. As the tenth son of his father it is said that he was considered at one time as a tithe for the ministry, but after one year in school his father withdrew him. Then followed a short time in a school where he learned writing, bookkeeping and navigation, before his father withdrew him again, at age 10, to start work with himself as a maker of soap and candles. Because this work did not please him he became apprenticed as a printer to an older brother for a number of years. This brother had returned from London with a printing press in 1717 and had started a local newspaper, *New England Courier*, in 1721. Quarrels with the local authorities, and between the brothers meant separation in 1723, and Benjamin was forced to look for a job elsewhere. This he found in Philadelphia, along with accommodation in a family whose daughter he was later to marry. He then moved to London after being financially tricked by the State Governor of Pennsylvania, but returned to Philadelphia in 1726 and soon had his own printing business shared with a partner. They published the *Pennsylvania Gazette* which Franklin kept when the partnership was dissolved in 1730. In 1733 he began the publication of *Poor Richard*, a collection of proverbs and maxims from around the world, which appeared annually until 1757. Four examples are:

> "You will be careful, if you are wise,
> How you touch men's religion, or credit or eyes."

> "Fear not death, for the sooner we die,
> The longer shall we be immortal."

> "Think of three things—Whence you came
> Where you are going, and to Whom you must account."

"Doing an injury puts you below your enemy;
Revenging one, makes you but even with him;
Forgiving, it sets you above him."

Franklin's Introduction to Electricity

Franklin's electrical interests may be said to have begun in 1743 when he saw a demonstration in Boston by a Dr Spencer from Scotland. Franklin was so impressed that he sponsored a similar demonstration in Philadelphia. The demonstration included suspending a small boy and causing sparks from his head and face by rubbing a glass tube at his feet. When that demonstration was over he purchased the whole equipment from Spencer. The following year Franklin received from Peter Collinson F.R.S. in London a similar glass tube with a copy of the *Gentleman's Magazine* for April 1745. This magazine reported the experiments of Hauksbee, Gray and DuFay with sufficient details to allow newcomers to repeat them successfully. This information to Franklin he shared with three associates, a silversmith, a lawyer and a teacher-preacher. By this time in his career Franklin had reached the age of 40 and had given up his daily printing business. For some time hereafter he was to have the opportunity to carry out experiments himself, which he was manually well able to do. But this opportunity was to last only until about 1753 when public service and diplomacy gradually used up all his time.

Franklin's Political Activities

In 1751 Franklin had been appointed a representative to serve in the Assembly for the city of Philadelphia. By the end of 1753 he was an important member of that Assembly, helped no doubt by his recognition abroad. He had been awarded the Copley Medal by the Royal Society in that year. At home also he had been awarded honorary degrees by Harvard and Yale, and had been appointed Deputy Postmaster general for all the colonies. As an important member of the Assembly he had close contact with the State Governor appointed by England. It was perhaps only to be expected that friction would develop between the Assembly and Governor, particularly over financial matters. The Assembly then appointed Benjamin Franklin as their Colonial Agent in London, expecting that his mission of representing the colony's views could be finished within six months. But since the Prime Minister in England refused to see Franklin, no doubt believing that it would be sufficient for his Crown ministers to see him, his visit lasted for between three and four years.

Franklin's Visit to Scotland

In the summer of 1759, two years after his arrival in England, Franklin decided that he would like to visit Scotland. In February of that year St. Andrews University had conferred on him the honorary degree of Doctor of Laws, simply sending him letters written in English and Latin apprising him of this honour. The Diploma read:

"Conferred the Degree of Doctor in Laws on Mr Benjamin Franklin famous for his writings on Electricity and appoint his diploma to be given him gratis, the Clerk and Arch Beadle's dues to be paid by the Library Questar." (The Questor was the chief financial officer of the University.)

In return Franklin sent a printed volume of his own electrical experiments to the College Library.

Apart from this gesture by St Andrews many of his closest friends in England had Scottish antecedents. His next door neighbour in London was a native Caledonian, with whom a friendship developed to the extent that in 1782 the British Foreign Office sent him as an envoy to negotiate with Franklin peace with the revolted colonies. And when the visit to Scotland did take place in 1759 Franklin had the opportunity of meeting David Hume, Adam Smith, Lord Kames the philosopher raised to the bench, and Principal Robertson of Edinburgh University the historian, all in the one visit. The journey to Edinburgh took one month, on a road on which highwaymen might have been encountered. Within three days of their arrival in the city, which had a population of around fifty thousand, both Benjamin and his son William were given the Freedom of Edinburgh. On this Scottish journey Franklin visited the Universities of Edinburgh, Glasgow and St Andrews, and also Blair Drummond, where he planted trees. Blair Drummond House was the home of the enlightenment thinker and jurist Lord Kames who began the transformation of the carse area, turning it from being water laden into good agricultural land.

Franklin's Discovery of the Identity of Electricity and Lightning

The discovery which will always be associated with Franklin, even by non-technical people, will be that of lightning being an electrical phenomenon. Linked with that is his invention of the pointed rod lightning conductor, as a means of dissipating the awesome energy in the lightning. The frightening awareness of this energy was known probably from time immemorial. Gibbon in his *Decline and Fall of the Roman Empire* tells the story of the Tuscan magicians in A.D. 408 who offered to draw lightning from the sky to destroy Alaric the Goth who with his hordes of barbarians were laying siege to Rome. Bishop Innocent counselled trying the experiment, but the Senate refused and Rome fell. With regard to the experimental proof that lightning was an electrical phenomenon Franklin himself carried out the experiment in June 1752. This was carried out with a kite because all buildings in Philadelphia were of single story construction, and no church existing there at that time had a steeple. Franklin's care in carrying out the experiment was presumably not equalled by Georg W. Richmann, Professor of Physics at St Petersburg, who tragically lost his life in investigating the electricity of thunderclouds in the following year. Richmann had been investigating electricity for some years, and had for the time a sound theoretical knowledge of the subject, coupled with what has been described by his successor as a first rate set of electrical apparatus. But the energy in lightning is orders of magnitude greater than that provided by the electrical apparatus of the time.

Franklin's Other Contributions to Electricity

Franklin's more general contribution to electricity was that it was a single, one fluid phenomenon. In this theory Franklin assumed that in each body there is a normal quantity of electric fluid. When that body is electrified it gains or loses some of its normal quantity and arrives at a plus (+ve) or minus (-ve) state. Associated with that there is a minus or plus state in the other body producing the electrification. This is an example of conservation of charge, that what is added to one body is taken away from another.

An example of this occurs in the Leyden Jar discovered by Musschenbroek (or possibly Von Kleist) in 1745. It was a device for both producing higher voltages and also storing the electricity produced by a rubbed glass tube. In its original form it consisted of hanging a wire from the prime conductor of the glass tube, and dipping the other end of this wire into a bottle of water held in one hand. When the operator standing on ground touched the prime conductor with his other hand he received a shock which shook him like a thunderbolt. His hand was earthed through his feet, and the capacitance between the inside of the glass and his hand had stored the cumulated charge from the glass tube over a period of time. The shock he experienced was described in a letter to a colleague in the following way: "I wish to communicate to you a new, but terrible, experiment that I would advise you never to attempt yourself."

Franklin gives two examples of the damage which the Leyden Jar was capable of producing. Using a nine-inch diameter globe and a half pint of water vial he was able after fifty rotations of the globe to kill hens outright. To stun a turkey of about ten pounds weight with the same globe he required two six-gallon glass jars of water and two thousand rotations of the globe. The build up to producing the accumulated output from the Leyden Jar he illustrated as follows:

"Suppose the common quantity of electricity in both sides of the glass bottle before the operation begins is denoted by 20. For every rotation of the globe suppose a quantity equal to 1 is thrown in. After the first rotation the quantity in the wire and upper part of the bottle is 21 and in the bottom only 19. After the second the upper part will have 22 and the lower 18, and so on, till after twenty rotations the upper part will have a quantity of 40, and the lower part none. The operation then ends, for if the rotations continue loud cracks will sound and sparks may fly out through the sides of the bottle. The convulsion to the body in the person holding the Leyden Jar is caused by the passage of the electricity from the inside to the outside of the vial. A communication with the floor is not necessary since he that

holds the bottle with one hand and touches the wire with the
other will be shocked even standing on wax."

In this way Franklin explained the operation of the Leyden Jar, which had
baffled European philosophers including its discoverer Musschenbroek. This
was so much the case that Musschenbroek said after discovering its effect
"he now understood nothing, and could explain nothing about electricity".
This degree of humility is rare, but he was a rare kind of man. It was said of
him that "he loved truth with the openness of a child". And when Franklin
wrote to him from America asking for his advice on what books he should
read on electricity, Musschenbroeh replied, "I would wish that you would go
on making experiments on your own initiative." He realised that Franklin
was a *rara avis* who was able to think independently of work which had been
done in Europe. He was, of course one of the two most eminent physicists in
Holland, born into a family of instrument makers which made air pumps,
microscopes and telescopes. Born in 1602 in Leiden, he obtained his degree
in medicine in 1715, and studied further in London where he attended
lectures by Isaac Newton. Thereafter he became Professor of Mathematics
and Philosophy in Duisberg in 1719, followed by a Chair in Astronomy at
Utrecht, and finally the Chair of Mathematics in Leiden. There he attracted
students from all over Europe and turned down offers of posts from the King
of Prussia and the Empress of Russia.

But Franklin's explanation of how the Leyden Jar operated was not
universally accepted.

F.U.T. Aepinus (1724-1802)

Franklin's theory of electricity was shown to be inadequate by the
mathematician F. U. T. Aepinus (1724-1802). Aepinus's father was Professor
of Theology at Rostock University in Germany, from 1721 to 1750. One of
his forefathers in turn had changed the German family surname from Hoch to
the Greek name Aepinus, because Hoch, meaning "tall or lofty", was similar
to, or possibly identical with, according to the spelling of the time, the Greek
word for "elected". This Lutheran forefather liked the association with the
reformed word "elected", and so changed the family name. Aepinus's father
had his sons taught by private tutors for their early education. At University,

parental pressure made F. U. T. Aepinus transfer to Medicine from his earlier registration in Mathematics. As part of his training he had to spend two years at Jena University where he studied Mathematics and Natural Philosophy as well as his Medical subjects. He finished at Jena in 1746, and was given a licence to teach Mathematics and Natural Philosophy in the following year. He also successfully defended his medical examination, but did not formally take his degree in that area.

Aepinus's Early Researches

F. U. T. Aepinus's research publications began in 1753 when he observed one of Mercury's relatively rare transits across the face of the sun. The following year he obtained the Chair of Astronomy in Berlin where the Swiss Mathematician Leonhard Euler was a dominant figure. Euler had moved there from St Petersburg in 1741 after being invited there by Frederick the Great. He was effectively Head of the Prussian Royal Academy of Science and Belles Lettres in Berlin, which owned eleven telescopes, and for the next two years Aepinus would have been in almost daily contact with Euler. He dined regularly at Euler's house with other researchers, including Euler's eldest son Johann Albrecht Euler who early in 1755 had been awarded the St Petersburg prize for an essay on Electricity. In the same year John Carl Wilcke arrived from Sweden, the son of a pastor of the German church in Stockholm, who hoped that his son would become a Minister like himself. But John Carl who shared a house with Aepinus transferred to Science, and became linked with Aepinus's later researches in electricity. Euler himself was a devout Calvinist and viewed the Leibnitz view of matter possessing an internal dynamics of its own as unacceptable theologically, and mistaken scientifically.

Possibly towards the end of 1756 the mineralogist Lehmann drew Aepinus's attention to the mineral tourmaline. This had the power to draw ashes to itself when placed on burning coal. Aepinus found that it had become electrified in this way. Moreover it acquired opposite charges on its two faces, instead of having one charge along its length. In this way it paralleled the operation of a magnetised piece of iron. The implication of this was that a theory for electricity might well be applied to magnetism as well. We will consider the electrical case only.

Aepinus's Major Contribution to Electricity

Aepinus's work was based on the earlier experimental discovery by Gray that electric charge, or as he put it electric fluid, was confined to a thin layer on the surface of a conductor. Hence attraction and repulsion between charged bodies must act at a distance across the intervening air. From Franklin's experiments he also accepted the following four propositions:

(1) Electric charges mutually repel each other.
(2) Electric charges are attracted by particles of ordinary matter. (Here it is helpful to think of ordinary matter as atomic nuclei with positive charges.)
(3) Electric charges move easily in conductors and with difficulty in insulators.
(4) Electric phenomena can involve motion of electric charges, or attraction or repulsion between static charges.

He begins by considering the forces on a charge by a piece of ordinary matter, containing within it other charges. Equilibrium will be established when the attractive forces are equal to the repulsive forces. This gives what Franklin called the "natural quantity" for the charges in this matter. If a body contains any other quantity than the natural it will be unstable. Applying this to a conductor placed in the vicinity of an electrified body, the nearer portion of the conductor will acquire an opposite charge to that of the electrified body, and the farther away portion will have the same kind of charge. It was possible to make this deduction without specifying the magnitude of these forces, since at that time the force between two electric charges was not known quantitatively as a function of their separation.

This deficiency is claimed to have been overcome by Priestley who experimentally verified in 1766/67 that when a hollow metallic vessel is electrified there is no charge on the inner surface. He then proceeds: "May we not infer, that the attraction of electricity is subject to the same laws as the attraction of gravitation; since it is easily demonstrated that were the earth in the form of a shell, a body on the inside of it would not be attracted to one side rather than the other?"

This has been described by Whittaker as "a brilliant inference". It does, however, raise the question of whether Whittaker was favouring an English natural philosopher, despite Priestley's own claim that he thinks he has kept clear of any mean partiality towards his own countrymen. It is fair to say that Whittaker goes on to credit John Robison of Edinburgh University with proving the law of force by direct experiment. Accordingly we will now consider John Robison and his contributions as one of the Franklin School.

John Robison (1739-1805)

John Robison was the son of a wealthy Scottish merchant who wished his son to become a Minister. He attended Glasgow Grammar School and then studied at Glasgow University, where Joseph Black was Professor of Chemistry, and at the same time James Watt was an instrument maker there. Robison was particularly influenced by three of his Professors, Robert Simson the mathematician, the classicist James Moore, and the natural philosopher Robert Dick. At the age of eighteen he published a better design for the Newcomen engine, and later in life he was responsible for the invention of the siren. After graduation Robison remained in Glasgow for a time, and became friendly with James Watt who was involved, through his position at the University, with a collection of astronomical instruments which had been bequeathed by Dr Macfarlane of Jamaica.

Robison as Tutor in Scotland, England and Quebec

Thereafter Robison became a tutor in both Scotland and England initially. In Scotland in a time of very intense frost he was living in a room in which the window was very badly fitted, so that more cold was admitted than was agreeable. He remedied this by applying water with a brush to the bad joints. When the water froze the joints were stopped. In England he became a tutor to the son of Admiral Sir Charles Knowles, who also became his Sponsor. As his son's tutor he was accorded the rank of midshipman when his charge sailed under the command of General Wolfe in the taking of Quebec. In the scaling of the heights by the Scottish Highlanders it is believed that the French sentries were confused by the Scottish Gaelic being mistaken for Brettonese. It is recorded that while stationed on the St Lawrence Seaway Robison put his mathematical knowledge to use by surveying the area.

Robison as Board of Longitude Assessor

When Robison returned to England he left the navy, but later accepted an offer in 1762 through the same Sponsor to test John Harrison's chronometer on its test run across the Atlantic to Jamaica. This was the first successful instrument of its kind. The following year he abandoned the sea, and there followed some years of uncertainty as he studied theology, was offered a Government Commission to survey North America and was also offered the opportunity of taking Holy Orders.

Robison in Russia (1769-1774)

In 1769 he chose to go to Russia to be Private Secretary to Sir Charles Knowles who had been appointed by Empress Catherine the Great to update her navy for a war with the Turks. By 1772 Robison was sufficiently fluent in Russian to be given a Professorship in Mathematics at Kronshtadt, near Leningrad, and seems to have secured Catherine's patronage. But Joseph Black in Glasgow engineered Robison's return as Professor of Natural Philosophy in Edinburgh. This clearly disappointed Catherine the Great, who nevertheless granted him a pension when he resigned in 1774.

Robison in Edinburgh

In Edinburgh Robison began a broad range of lectures of high academic standard. These were not popular with the students, both because of the mathematical level required for their understanding, and because he refused to provide popular type demonstrations. Consequently he became one of the poorest paid of the Professors, since his income depended on the number of his students. However, outside the University he revitalised the Philosophical Society of Edinburgh, and when it became the Royal Society of Edinburgh in 1783, he became its General Secretary.

Throughout his life Robison maintained a close friendship with both Joseph Black and James Watt. After Black's death he decided to publish Black's Chemical Lectures. These he found to be in a disorganised state, but because he knew of their scientific merit he persevered with the work, which was published in 1803, four years after Black died.

With regard to the relationship with James Watt this friendship appears to have been even closer. Hundreds of letters testify to this. In 1796/7 Robison testified in London to the validity of Watt's invention of the separate condenser for the steam engine, and this intervention had a very great effect upon the judge and jury. To give this evidence Robison had to obtain leave of absence from Edinburgh University.

Later in life Robison wrote a lengthy book entitled *Proof of a Conspiracy Against all the Religions and Governments of Europe*. This conspiracy he asserted was carried on in secret meetings of Free Masons and others, which had resulted in the French Revolution of 1789. He himself was a Free Mason, even while he lived abroad. The book was controversial, but did not prevent the award of honorary degrees to him by Princeton University in 1798, and Glasgow University in 1799.

Joseph Priestley (1733-1804)

The third edition of Joseph Priestley's book on *The History and Present State of Electricity* was published in 1767. He was elected to the Royal Society in 1766, and received the Copley Medal in 1773, although this was for his discovery of oxygen. Priestley (1733-1804) was both a theologian and a scientist. He was born in Yorkshire, the son of a strict Calvinist. His rebellion against the established churches resulted in him becoming a dissenting Minister at age twenty-two. But after five years he became a Classics teacher in Warrington, and earned a doctorate from Edinburgh University for his work on Education. Having met Benjamin Franklin in London he wrote his *History of Electricity*, with the help of Franklin, Watson and Canton, who were all experimenters of repute themselves. Consequently his work was an up to date description of qualitative electricity for the time. But it lacked quantitative data. Throughout the two volumes there is not a single equation. Priestley was not skilled in Mathematics. Moreover although he claims to have kept clear of any "mean partiality towards his own countrymen", and in his third Edition says he has given a fuller account of discoveries made by several foreigners, it is a sad omission that he has apparently not attempted to assess the original work of F. U. T. Aepinus. Although this was originally written in Russian it would be surprising if Priestley were not aware that the Chair of Natural Philosophy in Edinburgh was occupied by John Robison, who was well aware of Aepinus's work in Electricity.

Franklin as Inventor, Philanthropist and Christian Deist

Franklin was responsible for many inventions. Among these were the rocking chair, a flexible catheter, bifocal glasses and the Pennsylvania fireplace. His idea of social responsibility, however, was such that he refused to patent any of his inventions, because he believed that their benefits should be freely available to all.

Franklin's social responsibilities also extended into the medical field. The idea of a Pennsylvania Hospital came from Thomas Bond, who was the physician among the first ten members of the American Philosophical Society. Bond wanted a hospital for the sick and insane, but met with only small success. Franklin was not an expert in medicine, but in promotion. He got together a meeting of citizens to consider Bond's idea, and in 1755 the cornerstone was laid. The wording on the inscription was devised by Franklin himself as follows:

In the year of Christ
MDCCLV
George the Second happily reigning
(For he sought the happiness of his people)
Philadelphia flourishing
(For its inhabitants were public spirited)
This Building
By the bounty of the Government
And of many private persons
Was piously founded
For the relief of the sick and miserable
May the God of Mercies
Bless the Undertaking.

This has been described as the language of philanthropic deism. It is in keeping with the spirit of a letter written to the President of Yale towards the end of Franklin's life. In that letter he writes:

"You desire to know something of my religion. It is the first time I have been questioned about it. But I cannot take your curiosity amiss, and shall endeavour in a few words to gratify it. Here is my creed. I believe in one God, Creator of the universe. That He governs it by His Providence. That He ought to be worshipped. That the most acceptable service we render Him is doing good to His other children. That the soul of man is immortal, and will be treated with justice in another life respecting its conduct in this. These I take to be the fundamental principles of all sound religion, and I regard them as you do, in whatever sect I meet with them.

"As to Jesus of Nazareth, my opinion of who you particularly desire, I think the system of morals and His religion, as He left them to us, the best the world ever saw or is likely to see; but I apprehend it has received various corrupt changes, and I have, with most of the present Dissenters in England, some doubts as to His divinity; though it is a question I do not dogmatise upon, having never studied it, and think it needless to busy myself with it now, when I expect soon an opportunity of knowing the truth with less trouble. I see no harm, however, in its being believed, if that belief has the good consequence, as probably it has, of making his doctrines more respected and better observed; especially as I do not perceive that the Supreme being takes it amiss, by distinguishing the unbelievers in His government of the world with any particular marks of His displeasure.

"I shall only add respecting myself, that, having experienced the goodness of that Being in conducting me prosperously through a long life, I have no doubt of its continuance in the next, without the smallest conceit of meriting it—I confide that you will not expose me to criticism and censure by publishing any part of this communication to you. I have ever let others enjoy their religious sentiments, without reflecting on them for those that appeared to me unsupportable and even absurd. All sects here, and we have a great variety, have experienced my good will in assisting them with subscriptions for building their new places of worship; and as I never opposed any of their doctrines, I hope to go out of the world in peace with them all."

These are the religious views of Franklin written towards the end of his life. But almost contemporary with Franklin there lived a man whose life could hardly have been more different in the way he lived it. This was Leonhard Euler, whom we will now consider.

Leonhard Euler (1707-1783)

Leonhard Euler was the son of Paul Euler who was a Calvinistic Minister in a small village near Basle in Switzerland. He himself had studied Mathematics under James Bernouilli, a family which produced eight distinguished men of science. This particular member of the family impressed on his pupils the importance in particular of geometry, with its manifold applications in the real world. Paul Euler taught his son Leonhard elementary mathematics to such effect, in addition to his innate ability, that when Leonhard entered Basle University he received additional private weekly instruction from James Bernouilli himself. This privilege in directing forward progress, and preventing the entering of dead-end pathways, justifies one-to-one teaching.

But after graduation the father's desire that Leonhard should study theology was the cause of a temporary rift until a common mind of continuing with mathematics was agreed. Thereafter Leonhard, at age 19/20, submitted an entry for the 1727 Grand Prize of the Paris Academy on the best arrangement of masts in ships. This essay won him second place, an astonishing achievement for one who had been brought up in a landlocked country. The first prize went to a Professor of Hydrography.

Euler in the Russian Navy (1727-1730)

An application by Leonhard Euler for a vacant Chair in the University of Basle was unsuccessful, but through the good offices of Daniel and Nickolaus Bernoulli who had gone to St Petersburg two years before, he was offered an appointment there. This was to teach the applications of mathematics and mechanics to physiology. Taking advantage of the time until the work had to be started, Leonhard Euler studied this subject and arrived in St Petersburg in May 1727. But on the day he arrived Empress Catherine I who had set up the Academy of Sciences died, and his job offer became unavailable. Nevertheless, Euler was offered a post as a medical lieutenant in the Russian navy, and he served there from 1727 to 1730.

Euler as Professor of Physics and Mathematics in St Petersburg (1733-1741)

In 1730 Euler became Professor of Physics in the Academy of Sciences, and when Daniel Bernoulli gave up the post of Professor of Mathematics in the Academy in 1733, it was given to Leonhard Euler. By 1740 he had a very high reputation, after having won in 1738 and 1740 the Grand Prize of the Paris Academy. Consequently when Frederick the Great invited him to go to Berlin where an Academy of Sciences was to replace the Society of Sciences he accepted and moved there in 1741. And there he stayed for twenty-five years.

Euler in Berlin (1741-1766)

The relation between Euler and Frederick the Great was not always harmonious. While Euler thought Frederick pretentious, he in turn thought Euler lacking in sophistication. The story is told that Euler was reproached by the Queen of Prussia for not conversing, to which he replied, "Madame, I come from a country, where if you speak you are hanged." No doubt he had in mind the reign of Ernest John van Biren, Duke of Courland, in Russia, of whom it is said that between 1737 and 1741 he caused 11,000 to be put to death and double that number to be exiled.

Euler's Breadth of Interests

The contributions made by Euler to Natural Philosophy were wide ranging. They include his Treatise on Mechanics, his calculations relating to establishing a reversionary fund, to secure to widows or orphans either a fixed sum or an annual revenue payable after the death of a husband or father, calculations on coinage, on forming navigable canals, and after he lost his sight the making of artificial magnets. He did not read the modern poets, but knew the Aeneid by heart.

In 1766 Euler returned to St Petersburg, where in 1771 the city suffered from a conflagration which reached his house. A native of Basle, knowing of his blindness, burst through the flames and carries Euler on his shoulders to safety. The house, which had been a present from the Empress, was replaced through her munificence again.

Euler's Religious Beliefs

As long as his sight remained he read to his family a portion of Scripture every evening. He was a Calvinist, the religion of Switzerland, his home country. But because the word Calvinist has acquired a disparaging meaning in today's usage, it may be helpful to quote some of his beliefs as follows. First with regard to creation:

Euler's Beliefs with regard to the Physical World

With regard to the physical world Euler believed

(1) The world corresponds to the plan which God proposed and implemented.

(2) Consider the eye. It is perfectly adapted to the purpose of representing distinctly exterior objects. And to keep itself in that state for a lifetime through the juices which preserve it.

(3) The same perfection is demonstrated in plants with regard to their growth, the production of their flowers, their fruits and seeds.

(4) Consider an oak tree springing from an acorn with the help of the food supplied by the earth.

Euler's Beliefs with regard to the Spiritual World

With regard to spirituality:

(1) The wickedness of man seems to be an infringement of this perfection.

(2) God is supremely good and holy, is the author of the world, which nevertheless swarms with crimes and calamities.

(3) The difficulty of reconciling the perfection of God with the existence of evil disappears if we remember the liberty given to the spirit of man.

(4) At the moment of creation spirits were all good, for time is required for the formation of evil formulations.

(5) The essence of spirits is freedom, including freedom to sin. This is not inconsistent with divine perfection.

(6) The wickedness of some men sometimes leads to the correction of others, and thereby leads them to happiness.

(7) All events which come to pass are always under the direction of Providence, and finally end in our true happiness.

(8) When a man addresses to God a prayer worthy of being heard, it must not be imagined that such a prayer came not to the knowledge of God till the moment it was formed. That prayer was clearly heard from all eternity. And if the Father of Mercies deemed it worthy of being answered, He arranged the world expressly in favour of that prayer, so that the accomplishment should be a consequence of the natural course of events. It is thus that God answers the prayers of men, without working a miracle. Thus all our prayers have been already presented at the throne of the Almighty, and have been admitted into the plan of the Universe, in subservience to the infinite wisdom of the Creator.

(9) God created the material world, with so many great works, with intelligent beings, who were capable of admiring it, and being elevated by it, to the admiration of God, and with the capability of the most intimate union with their Creator, in which their highest happiness consists.

(10) Hence intelligent beings, and their Salvation, must have been the principal object of the Creation. All the events in Creation are in

harmony with the wants of all intelligent beings to conduct them to their true happiness.

(11) Every man may rest assured that, from all Eternity, he entered into the plan of the Universe. This consideration must increase our confidence, and our joy in the Providence of God, on which all religion is founded.

(12) Liberty is essential to every spiritual being. As God cannot divest a body of its space and inertia, without annihilating it, so He cannot divest a spiritual being of its Liberty. But God can persuade people by providing us with motives which His Providence supplies, e.g. a man may be affected by a sermon and be converted. This is the result of Divine Grace conducting intelligent beings to happiness and salvation.

(13) The physical characteristics of bodies are that they occupy space, they possess inertia and they cannot penetrate solids. On the other hand spirits occupy no space, they possess no inertia and they can penetrate solids. Because of them occupying no space the mediaeval philosophers raised the question of how many angels could dance on the point of a needle, which is not relevant to their purpose.

(14) My soul does not exist in a particular place, but it acts there.

(15) God is everywhere but His existence is attached to no place.

(16) Every spirit lives.

(17) A body by itself has no intelligence, no will or liberty.

(18) It is the spirit which produces the illustrious actions of intelligent beings. This power which every soul has over its own body cannot be considered but as a gift of God, who has established this wonderful union between soul and body.

(19) The influence of the soul upon the body constitutes its life, which continues as long as this union exists. Death is the dissolution of this union. At death the soul, since it resides in no place, has no need to be transported anywhere. Since God is an infinite Spirit He is everywhere present.

(20) Sleep furnishes us with something like an example of the separation of soul and body, with the soul providing dreams, but it is not possible to say much about these.

Euler's Contributions to Electricity and Magnetism

Euler was a consummate mathematician. What he said could be checked according to the established rules of mathematics. He was also required to be a natural philosopher, since Frederick the Great asked that he provide his niece Princess of Anhalt Dessau with lessons on elementary science. This resulted in a three volume set of 234 letters, entitled *Letters to a German Princess*, which are a masterpiece of exposition. It included thirteen letters on electricity and eighteen letters on magnetism. The information in these letters was necessarily based on experimental results which had been accumulated over hundreds of years. A few of these were checked by Euler himself, but he was influenced by the contemporary view of an all-pervading aether which could pass through the pores of all objects with varying degrees of difficulty.

Euler's Views on Electricity

With regard to electricity he was aware of and accepted the classification of Franklin that a body can become electrified in two different ways. That is, one in which the aether becomes more compressed, and this is denoted by positive electricity, and the other in which it becomes more rarefied, denoted by negative electricity. Euler was also aware of the effect of bringing one's face near an electrified body. This gave a feeling similar to the application of a spider's web on the face. Likewise he was familiar with the method of producing higher strength electricity by rotation of a glass globe, and how this could be used to set alcohol on fire, and also to kill birds by means of an electric shock. With regard to the application of electricity to humans he knew that it had been observed that the pulse of an electrified man beats faster, and he "is thrown into a sweat, though sufficient experiments have not yet been carried out to know in what cases this effect is salutary or otherwise". Finally he was aware that when a finger is presented to a conductor with positive electricity so that a spark is produced, the geometric form of the light rays are different for the positive and negatively charged conductors.

Euler's Views on Magnetism

With regard to magnetism he refers to a large loadstone which was so powerful that at a distance of several feet a mass of iron continued to experience a considerable force. Reference is also made to Mahomet's coffin, supported, it is said, by a loadstone. On the grounds that artificial magnets had already been constructed which carry more than one hundred pounds weight, he suggests that this is not impossible. This appears to assume that the technology for making such artificial magnets, or better, was available at the appropriate time.

At the time that Euler wrote the letters to the German Princess in 1760/61 the force between electric charges or between magnetic poles was not known quantitatively as a function of their separation. This was discovered for electric charges by John Robison of Edinburgh in 1769, who found it to be an inverse square law—so that doubling the separation resulted in a reduction of the force by a factor of four. Then just two years after Robison's discovery a memoir was presented to the Royal Society by Henry Cavendish, an Englishman born in Nice, who merits a section of his own.

Henry Cavendish (1731-1810)

Henry Cavendish came of a titled family, and was the brother of the third Duke of Devonshire. At age eighteen he entered Peterhouse College, Cambridge, and left four years later without taking a degree. During his lifetime family money was left to him to the extent that when he died he left more than £1,170,000. His mother died when he was only two years old and he never married. He lived and worked alone. It has been said of him that he was the richest of the philosophers and the most philosophical of the rich. But he did attend meetings of the Royal Society regularly. It has been reported that new Fellows of the Royal Society who wanted to talk to him were given these instructions: "Never look at him, but talk as it were into a vacancy, and then it is not unlikely but you may set him going." When he was set going he talked to the purpose and yet could broaden the discussion with his range of knowledge. His shyness of women was such that it is said that he had a back staircase added to his house in order to avoid encountering his housekeeper.

It has been said above that Cavendish sent a memoir to the Royal Society in 1771. This was entitled, "Attempt to explain some of the principal phenomena of Electricity by means of an elastic fluid." This was similar in concept to the earlier work of Aepinus, but discovered independently. If we ask what it would have been possible to measure in 1771 we have to recall that there were no measuring instruments apart from the pith ball electroscope for comparing the strength of electric charges. The current words in electricity were fluids and atmospheres. Words such as current, voltage, resistance, capacitance or inductance had not been invented.

Cavendish's contribution was to define concepts and verify them by careful measurement. He defined charge and degree of electrification (which we call potential difference). He defined also capacitance, resistance and current density. He also verified the inverse square law experimentally showing that the force must vary inversely as 2 plus or minus 1/50. But most of his research results were not published until 1897, after Clerk Maxwell agreed to edit his notebooks, following their discovery by William Thomson in 1849. Thomson had discovered that Cavendish had measured the capacitance of a circular disc as 1/1.57 of that of a sphere of the same radius. And the equipment available to Cavendish consisted of a pith-ball electroscope and a friction machine generator. Further experiments by Cavendish included a comparison of the conducting powers of different materials. Thus in a memoir to the Royal Society in 1775 he asserted that "iron wire conducts about 400 million times better than rain or distilled water—that is, the electricity meets with no more resistance in passing through a piece of wire 400,000,000 inches long than through a column of water of the same diameter only one inch long. Sea-water, or a solution of one part of salt in 30 of water, conducts 100 times or a saturated solution of sea-salt about 720 times, better than rain water." In the detailed account of the experiments published in 1879 the method of testing appears to be that of physiological sensation alone.

The first person to assert that magnetic poles interact according to the law of inverse squares was John Michell, at one time a Fellow of Queen's College, Cambridge, and later an Anglican priest. He discovered a good way of making artificial magnets, which were less expensive than natural loadstone, because the cost was little more than the steel they were made of. They were also stronger magnetically, were more easily restored after damage and could be of any shape. He was also the inventor of the torsion balance, a means of measuring the twisting of wires, which he then proposed as a means of measuring the very small force between magnetic poles. A piece of equipment to carry this out was constructed before his death in 1793, after which it eventually passed to Cavendish. In the intervening time a wooden arm had become warped and Cavendish decided to place the whole apparatus in a room kept constantly shut, viewing the movements through a telescope. Before this was reported, the French scientist C. A. Coulomb had used the torsion balance for measuring the force between electric charges in 1785, but Cavendish states that Michell had described the torsion balance to himself before that date. The result is that the names associated with the forces between charges, and between poles are sometimes linked with the different names of Michell, Robison, Cavendish and also Coulomb. The work of Coulomb will now be considered.

Another inventor of artificial magnets was Gowin Knight, a clergyman's son who went to the same school as John Smeaton, and was the first person to take out a patent in magnetism.

Charles Augustin Coulomb (1736-1806)

Charles Augustin Coulomb was born in 1736 in Angouleme, about seventy miles North East of Bordeaux. His parents originally wished him to study medicine, but he rebelled and his mother disinherited him. It appears that he and his father then moved to Montpellier, west of Marseilles, after his father suffered from speculative financial losses. He studied mathematics there from 1757 to 1759, before returning to Paris to prepare for entry to the engineering school at Mezieres. This he entered in 1760 and graduated the following year with the rank of lieutenant en premier in the Engineering Corps. He was posted to Brest initially, but in 1764 was transferred to Martinique in the West Indies. There he was involved in the construction of Fort Bourbon, directing several hundred labourers, and acquiring experience for his later researches in mechanics. After eight years he returned to France in 1772 with the rank of Captain, and presented a memoir to the Paris Academy in 1773 on the results of his engineering experience. He then decided to compete in the Academy's competition for 1777 which required improved compass needles and an explanation of the daily variation of the earth's magnetic field. His submission shared the first prize and set him off in this subject area which was completely new for him.

Before an explanation of the variation in direction of the earth's magnetic field could be given it was necessary to establish how much that variation was. The existing method of measurement used a pivoted magnetic needle in which the frictional force might vary as much as the variation it was designed to measure. But Coulomb had experience of twisting in ropes, and implemented first a fine silk thread for the suspension of the needle. But this proved too sensitive, and was replaced by a wire. Within certain limits Coulomb found that the torque exerted by a twisted wire was proportional to the fourth power of its radius, and to its angle of twist, divided by the length of the wire. This enabled Coulomb to obtain results for the force between charges which he doubtless believed were new, but agreed with those which Robison had previously obtained.

The law for these charges is now known as Coulomb's Law. He has also been honoured by the definition of the coulomb as the quantity of electricity passed by one ampere in one second. Further Coulomb has had his name inscribed on the Eiffel Tower. Further discussion of his work will be continued in Chapter 4.

Summary of Chapter Three

Covering the period approximately 1740 to 1800 contributions to the study of electricity and magnetism were made on both a European front, and more unexpectedly on a North American front. Van Guericke in Germany demonstrated both attraction and repulsion using a spherical globe of sulphur which had been manufactured in a glass envelope. Isaac Newton showed that the sulphur was unnecessary, and that a glass globe alone was sufficient to produce electricity. Hauksbee in England extended these measurements to evacuated globes and to glass rods and tubes and used a suspended boy to demonstrate sparks coming from him. Bose in Germany was even more of a showman, but contributed the idea of a metal tube from the far end of which it was possible to combust alcohol and other materials. Watson in England was a Merchant Taylors's boy who extended an electrical circuit to a distance of four miles to find the velocity of propagation. He was also responsible for introducing Franklin in America to the Royal Society.

Franklin was responsible for introducing the idea of positive and negative electricity, and for confirming that lightning was electricity. Aepinus in Germany and Robison in Scotland emphasised the importance of theoretical understanding of electric charges at rest and in motion. Priestley in England provided a long non-mathematical survey of electricity covering two volumes. Euler from Switzerland worked in Russia and Germany, providing a clear summary of subjects in natural philosophy for a German Princess. Cavendish in England and Coulomb in France both contributed to a more complete theoretical understanding of forces between charges and poles.

CHAPTER FOUR

APHORISMS OF LICHTENBERG

(c.1740 – c.1800) (SECTION 2)

THIS CONTINUATION of Chapter Three deals first with the achievements of G. C. Lichtenberg (1741-1799) at the University of Gottingen in Germany, then with Count A. C. Volta (1745-1827) at the University of Pavia in Italy, followed by Charles Augustin Coulomb (1736-1806) in Martinique in the West Indies and also in Paris. This French connection is continued with the Marquis P. S. Laplace (1749-1827), Professor at the Ecole Militaire in Paris and finally with his pupil Simeon-Denis Poisson of the Ecole Polytechnique there.

In this group Lichtenberg may be described as a polymath, achieving fame in astronomy, physics, geometry, meteorology, and chemistry, and in devising the concepts of repression, compensation and sublimation in the subconscious, and also in producing a book of aphorisms. Nevertheless despite such achievements his name is little known to the public, by comparison with that of Volta, who invented the electric battery, and has had his name attached to the electrical unit of potential difference across the terminals of such a battery. To be fair to Volta, however, he was also responsible for the invention of a device which was the first static electricity generator. He called it an electrophoresis, meaning a carrier of electricity, and a description of this will be given in a later section.

Coulomb chose to become a military engineer rather than a civil engineer where his work would have been restricted to building roads and bridges. At that time the School of Military Engineering at Meziere was reputed to be the best such Technical School in Europe. Three days of the week were set aside for theory, with half of this being devoted to Mathematics. The other three days were spent on practical surveying and field mapping, and might include the construction of bridges and arches for the local community.

Laplace was primarily a theoretician in contrast to the other three. His dominant contribution was one he took from his gravitational research. In these he had defined the gravitational potential at a point distant "r" from a mass M as (M/r), and associated with this was a gravitational force which was equal to the derivative of this, in the direction of "r". Transferring this to the electrostatic case in which there is a charge Q, he defined the electrostatic potential as (Q/r), and associated with this there is an electric force equal to the derivative of this potential, in the direction of "r". Two useful outcomes followed from this definition. The first was that in a situation where many charges existed the total potential was simply the arithmetic sum of the individual potentials due to the separate charges. The second advantage was that on the surface of any conductor the potential was inherently constant, since otherwise the charges would move and the problem would no longer be an electrostatic one.

Poisson was a mathematician, to the extent that the only aphorism with which he is credited is the saying, "Life is good for only two things, discovering mathematics and teaching mathematics." Like a number of other people he was clumsy with his hands, but he compensated for this by his extraordinary gifts in mathematics.

Each of these five contributors to the increased understanding of electricity will now be considered in turn.

Georg Christoph Lichtenberg (1742-1799)

Georg Christopher Lichtenberg was the son of Johann Conrad Lichtenberg, a Protestant Minister in a small town near Darmstadt in Germany. His grandfather was originally a tutor to a family listed among the nobility, but gave up this post to become a pastor in a Pietist church after his religious conversion. Georg Christopher himself was educated at home initially, but after his father died when Georg was only ten, he transferred to the local school in Darmstadt. There he stayed for nine years, repeating the top class three times because his mother was unable to pay the fees for entry to Gottingen University. A scholarship was eventually provided by the Count of Hesse-Darmstadt which allowed him to enter the University at age twenty-one. However a condition was attached to the offer that he would devote himself entirely to philosophy, and especially to higher mathematics. And in addition he would eventually have to take up the post of Professor of Mathematics at Giessen, a smaller town about forty miles north of Darmstadt, though in fact he did not do this.

Lichtenberg as a Student

Lichtenberg's mother died within a year of his going to Darmstadt, and he never returned there. But he did retain contact with the other members of his family, and possibly his gratitude to the Count may have been responsible for his lifelong custom of calling the Duchy of Hesse-Darmstadt "my Fatherland". However he was also an Anglophile, and his only distant journeys away from his country in the whole of his lifetime were to England.

Lichtenberg studied at Gottingen for four years, benefiting from the teaching of A. G. Kastner, the Professor of Mathematics and author of books of both school mathematics and the History of Mathematics. He was a Fellow of the Royal Society and has a lunar crater named after him.

As a student in Gottingen Lichtenberg performed his academic work well, not just in natural science and mathematics, but also in modern languages. In particular at the end of his course he tutored young Englishmen who were studying at Gottingen, and seemed to admire their openness of character. It was also during these early years that he began his lifelong habit of writing his *Waste-Books*, his collection of thoughts, witticisms, quotations, and aphorisms for which he is famous.

Lichtenberg in England

His visits to England resulted initially from an invitation by the father of one of his English pupils to accompany his son to London, and the first visit there lasted for four weeks in 1770. During the course of it he met King George III, who was also Elector of Hanover, and who had inherited a collection of scientific instruments from his uncle, the Duke of Cumberland. Part of the reason for his collection, in turn, may have been his desire to obtain a geographical survey in particular of the Highlands of Scotland, for which land surveying instruments would have been required. But in addition, George III had been tutored by the Earl of Bute, who was born in Scotland in 1713, and had an interest in natural philosophy. He later became Prime Minister, in 1762, but only for one year, retiring to pursue his interest in science, particularly botany. But his interest in science resulted in his earlier acquiring his own personal collection of scientific instruments. And when George III became King in 1760 he too took steps to build up his own collection, which is now in the Science Museum in London.

After Lichtenberg's return to Gottingen in 1770, he undertook astronomical and geographic work to determine the positions of several towns in Hanover and also lectured on Probability. His second visit to England took place between September 1774 and Christmas 1775, and this included being a guest of King George III at Kew. There is also a reference, during one of the visits to England, of the King visiting Lichtenberg in his hotel, which indicates his fame and the respect in which he was held. During his visit he attended debates on the American Colonies and many playhouse performances. His visit to Westminster Abbey reminded him of Psalm 90 as he saw the tombs of kings: "Before the mountains were brought forth, or ever you had formed the earth and the world, even from everlasting to everlasting you are God." He visited the Observatory which was being built in Oxford, which far surpassed his own in Gottingen. Among other acquaintances he met were members of Captain Cook's crew and Joseph Priestley, the non-conformist scientist.

Lichtenberg as Professor in Gottenberg

During his stay in England Lichtenberg had been appointed Professor at Gottenberg. His reputation as a lecturer was excellent and he was able to attract as many as a hundred students to his lectures, which were given in the house in which he stayed. The income from student fees added to his small University salary. Because of the large quantity of physical apparatus he collected for his demonstrations, the University purchased this on his retiral. His best known experiments were those demonstrating what became known as Lichtenberg Figures after him. These figures appeared on the surface of a dielectric, which after being physically rubbed had a voltage applied between its top surface and earth. The discovery of the patterns which appeared on the top surface of the dielectric in the presence of a dusty atmosphere was described by Lichtenberg himself as follows:

> "The dust did not settle regularly, but to my great delight formed little stars in several places, which were at first unclear and difficult to distinguish, but later, when I deliberately strewed on the dust more thickly, became clear and beautiful, resembling in places a design in relief. Countless little stars showed up here and there, whole Milky Ways and greater Suns; there were arches, indistinct on the concave side, but on the convex decorated with rays; then neat little branches, not unlike those which the frost forms on windowpanes; little clouds of many shapes and shadings; and finally many figures of special shapes. It was a pleasant spectacle for me to observe, particularly when I saw that they could hardly be broken up; for even when I carefully removed the dust with a feather or a piece of cat's fur, the same figures were formed anew and often even more beautifully than before."

This picturesque language is of a kind which is attractive even to non-scientists. And in his desire to publicise such a scientific result Lichtenberg proceeded to invent what has become known as xerography. It was obvious that the only way people would be able to see these stars and clouds would be for them to come into his laboratory. But Lichtenberg found that by pressing a clean sheet of paper on top of these dusty pictures that the dust transferred to the paper and thus he obtained a copy of that picture. So began the history of electrostatic copying machines, the further development of which lay dormant for almost two hundred years.

Lichtenberg Figures may also be produced naturally by lightning strikes on grassy surfaces such as golf courses. Alternatively they may be produced in the design process for high-voltage equipment using the electrical treeing which occurs before insulation breakdown. The shape of the Lichtenberg Figures depends on the position of the high-voltage point which is excited, where there is a single such point, or more generally on the excitation area of the dielectric surface.

Three years after the discovery of the Lichtenberg Figures the first successful installation of a lightning conductor in Gottenberg was carried out by Lichtenberg to protect the house in which he was staying. It appears that other people in Gottenberg had made unsuccessful attempts before then, by not grounding the lightning conductor.

As part of his teaching programme in dynamics Lichtenberg demonstrated the lightness of hydrogen gas in his lecture theatre in 1782 by filling pigs' bladders with this gas, and watching the bladders rise to the ceiling. It was the following year that the famous first flight of a balloon carrying a human being took place in Paris. Lichtenberg's comment on this achievement was: "This brings man nearer to heaven with his body now, as well as with his intellect."

In 1784 Lichtenberg was visited in Gottenberg by Volta, the inventor of the voltaic battery, about whom more will be said later. But in addition to Volta, others who visited him there included Goethe and William Herschell the astronomer, who discovered the planet Uranus and two of Saturn's satellites, and who also observed the phenomenon of its rings. So his fame had spread further. Also among his students in Gottenberg were Pfaff who became famous for his work on differential equations, Gauss who was pronounced by Laplace to be the greatest mathematician in Europe, and von Humboldt the traveller and naturalist. In the same year Lichtenberg himself considered the volume of lava ejected from Vesuvius during the particular eruption of 1784.

In 1786 he proposed the system of paper sizes which are almost universally used in offices throughout the world today, apart from those in North America. They became the official United Nations document size in 1975. What Lichtenberg did was to take a sheet of paper of area one square metre, and ask the question, "What should the length to width ratio of this sheet be in order to produce no wastage when the sheet is subdivided any number of times?" He worked out the answer that it should be the square root of two, i.e. 1.414.

In 1789 he married his wife of six years so as to ensure that she should have a pension when he died. In 1793 he became a Fellow of the Royal Society and was offered the Chair at Leiden, which had perhaps the highest reputation in Germany. This he declined, probably on the grounds of ill-health from which he suffered all his life, because he was a hunchback. But he was still able to write a biography of Copernicus in 1795, and died in 1799, the year that Volta invented the low voltage battery.

Lichtenberg's Literary Aphorisms

Lichtenberg's picturesque language as given in his description of the Lichtenberg Figures was not confined to science. It was the outward sign of how important the meanings of words were to him. And it found its further expression in the aphorisms which he confided to his *Waste-Books*. These were sufficiently numerous to be made available in book form after his death, and are the contributions for which he is most famous in Germany. They comprise both general and religious truths, and unlike Euler's, there are fewer religious ones. In all, Lichtenberg wrote more than a thousand aphorisms. Translations of some of these into English have been made by different authors. The following selection of thirty-five is divided into General and Religious.

General Aphorisms

(1) One must not judge people by their opinions, but by what these opinions make of them.

(2) A book is a mirror. If an ass peers into it, you can't expect an apostle to look out.

(3) A grave is still the best shelter against the storms of destiny.

(4) The most dangerous of all falsehoods is a slightly distorted truth.

(5) To read means to borrow. To create out of one's reading is paying off one's debts.

(6) I am always grieved when a man of real talent dies, for the world needs such men more than heaven does.

(7) It is the gift of employing all the vicissitudes of life to one's own advantage and to that of one's craft that a large part of genius consists.

(8) Hour glasses remind one not only of the passage of time, but also of the dust to which we shall come one day.

(9) I have looked through the register of diseases and have not found cares and sad thoughts among them. That is wrong.

(10) Any experiment through which new knowledge comes in a very concentrated form is the work of God suddenly revealed to man's intellect.

(11) I have wished myself dead, but only under the blanket, so that neither death nor man could hear it.

(12) What makes heaven so agreeable to the poor is the thought that there the social classes are more nearly equal.

(13) Of another person, "He had an air about him that the Pietists usually describe as an exalted state of being, the armchair teachers of theology as piety, the sensible man with a knowledge of the world as simple mindedness and lack of common sense."

(14) When you implore learned men for the love of heaven to take some most humble notion to heart, scarcely have they read it than it goes straight to their heads.

(15) Whoever has less than he desires must know that he has more than he is worth.

(16) I am extraordinarily sensitive to noise of every kind, but it entirely loses its unpleasant effects as soon as it is connected with a rational end.

(17) The world beyond the polished glass of telescope or microscope is more important than that beyond the seas, and is surpassed perhaps only by the one beyond the grave.

(18) In writing things down one notices a great deal which one is not aware of in mere meditation.

(19) There is a certain state of mind (which, at least with me is not unusual), when the presence and absence of the loved one, becomes equally difficult to bear: or when at any rate I do not find in her presence the pleasure which, to judge by the unbearableness of her absence, I had expected to find in it.

(20) Man loves company, if it is only that of a burning rushlight.

(21) Some people become savants as others become soldiers, merely because they are not fit for any other occupation.

(22) People who have read a great deal rarely make great discoveries. Invention presupposes an extensive independent contemplation of things. One should look rather than be told.

(23) A man never foregoes anything without expecting a return—hence the storing up of rewards in heaven.

(24) If there were only turnips and potatoes in the world, perhaps someone would come along and say "It's a pity that plants grow the wrong way up".

(25) The idea which we conceive of as a soul is in many ways similar to the idea of there being a magnet in the earth. It is a mere image.

Religious Aphorisms

(1)   The soul places the countenance round itself like a magnet does with iron filings.

(2)   Never undertake anything for which you wouldn't have the courage to ask the blessing of heaven.

(3)   He believed the words "Divine Service" should be reassigned and no longer used only for attending Church, but should include good deeds.

(4)   God created man in His own image says the Bible. Philosophers reverse this and create God in theirs.

(5)   There is something in our minds like sunshine and the weather, which is not under our control. When I write, the best things come to me from I know not where.

(6)   It has been observed long since that when the spirit rises up, it lets the body fall on its knees.

(7)   The world is a body common to all men, and changes in a part of it to bring about changes in the souls of all who happen to be facing that part.

(8)   The only time we admire God's omnipotence in a thunder storm is when there isn't one, or after it is all over.

(9)   What an odd situation the soul is in, when it reads an investigation about itself.

(10) With most people unbelief in one thing is founded upon blind belief in another.

Alessandro Volta (1745-1827)

Alessandro Volta, the youngest of seven children, was born in Como, about thirty miles north of Milan. His father had been a Jesuit for eleven years, and the family background was so closely associated with the Catholic Church that Alessandro himself, throughout his life, tended to choose clerics as his closest friends. During one period of his life he investigated the beliefs of the Jansenists, who wished to be a reforming force within the Catholic church, but who were later suppressed by the ecclesiastical authorities on doctrinal grounds. The influence of the church, together with the views of contemporary society, also influenced his marital position, so that eventually he did not marry the opera singer Marianna Paris, because of the inequality of their social positions. Later he did marry a socially acceptable partner at age forty-nine, and at this time provided a significant donation to the Paris family in lieu of marriage.

Volta's education was largely controlled by a paternal uncle from the age of seven, when his father died. This education initially began in the local Jesuit college, but when it appeared that attempts were being made to recruit him at age sixteen into the Jesuit fold, Alessandro was removed by a Dominican uncle. For the next few years he was able to study science under the guidance of a future Canon Gattoni, who had himself been trained by the Jesuits. Canon Gattoni was the first person to erect in 1768 a lightning rod conductor in Como. Although his uncle wished to make Alessandro an attorney, which was a profession associated with his mother's side, he himself preferred at age eighteen to study electricity.

His earliest results were obtained at age twenty. Como was, and still is, associated with silk. Volta discovered that silk rubbed by hand produced positive electricity, but when it was rubbed by glass it produced negative electricity. However, his first major invention was the device known as the Electrophorus at age thirty in 1775. The word Electrophorus means a carrier of electricity, with the suffix "phore" indicating the word "carrier", just as the word semaphore indicates a carrier of signs or signals. But the properties of this instrument which Volta had devised were such that Volta was able to describe it, not just as a carrier of electricity, but as a *perpetual* carrier of electricity. To see how this claim can be justified requires a description of its operation as follows:

The Electrophorus consists of two components:

(1) A metal plate on a dielectric surface
(2) A separate metal plate with an attached insulated handle.

In its operation the dielectric plate is charged by rubbing it, so as to leave it with, say, a positive charge on its surface. The separate metal plate with the insulated handle is then set on top of the dielectric surface, with the result that the bottom of the metal plate is charged negatively, and its top surface is charged positively. A contact is then made between this top surface and ground, so that the positive charge flows to ground, leaving the metal plate negatively charged. The insulating handle is then grasped and the metal plate is removed, carrying with it the negative charge. This negative charge can then be used to charge a capacitor or Leyden jar. The whole process is then repeated, without having to recharge the dielectric surface. Thus the charging process of the capacitor may be repeated again and again, so justifying Volta's description as a perpetual carrier of electricity.

Volta did not fail to recognise and also publicise the value of his electrophorus. He maintained an extensive international correspondence, supplemented by visits to foreign physicists. Thus in Paris in 1781/2 he met Franklin who was then the American Agent there, Laplace and Lavoisier. In London in 1782 he met Cavallo, the Neapolitan physicist who had settled there, and who was a prolific author of papers on electricity. Then in 1784 he was able to visit Gottingen. There he met Lichtenberg who had come to electricity through curiosity about the electrophorous. This had attracted him so much that he had built a huge apparatus with a fifty-one pound cake, with a diameter of several feet which produced enormous length sparks.

In his journeys abroad Volta took with him a travelling bag packed with instruments, which Lichtenberg said "lay scattered in my house for several days". Volta was a natural showman who was aptly described in one of Lichtenberg's aphorisms: "Volta is rich in knowledge and knows how to show it." But on his trips abroad he acquired information which was useful to his department in Pavia. Thus in Gottingen he saw that the Library was housed in the largest building in town, and was lavishly equipped. German scientific instruments were also assessed and ordered for Pavia with government funds provided by Lombardy.

A second invention which is to the credit of Volta is the Condenser Electroscope. This was described by him in a paper to the Royal Society in 1782, and for which he was awarded the Copley Medal in 1794 after his election as a Fellow in 1791. The Condenser Electroscope was designed to measure electric charges smaller than could be measured with a simple gold leaf electroscope. It consisted of two horizontal metal plates, which were lacquered to prevent them coming into conductive contact. The lower plate is connected to the rod carrying the gold leaves of the electroscope. An insulating handle is connected to the top plate. To measure a weak electric charge the electrified body is placed in conducting contact with the fixed plate. Then the top plate is placed above it and touched to earth. The charge which flows to the gold leaves is thus augmented by the large mutual capacitance between the two horizontal plates. This capacitance is large because of the thin layers of insulating shellac between them. It may be that this discovery by Volta was experimental, since it is known that in recharging the electrophorus he had previously tried connecting a Leyden Jar to the metal plate rather than to the dielectric surface. This was unsuccessful, but he is likely to have remembered this effect.

The third and best known invention of Volta is that of the dry battery. Because of its enormous social effect it will be considered in the following section.

The Dry Battery

Until 1796 Volta measured weak electricity using frogs or his tongue. The usefulness of frogs had been discovered by Galvani, who in 1794/5 had obtained contractions of frogs' legs by putting certain parts of frogs' bodies in contact with other parts. No metal was involved in the experiment. At that time no other electrometer was as sensitive as a frog.

In 1796 Volta had a new type of electrometer made by his instrument maker Giussepi Ra. This was based on a design by William Nicholson which had been published in the Philosophical Transactions in 1788. This design was an engineered version of Abraham Bennet's electric doubler of 1887, which had also been published in the Philosophical Transactions of the Royal Society. It is recorded that within two weeks of Ra's instrument being completed, Volta was able with this doubler to detect the electricity from two different metals which had been touched and then separated. Previous to this only frogs' legs and men's tongues had been sufficiently sensitive to do this.

Further work by William Nicholson again contributed to Volta's progress towards the development of the dry battery. In 1798 Nicholson wrote a paper on the Torpedo Fish, which in different species could produce a voltage of between eight volts and two hundred and twenty volts, sufficient in some cases to kill its prey. In this paper he stated that "the electric organs contained from five hundred to more than one thousand columns, about one inch long and one fifth of an inch in diameter, filled with many transverse films 0.0033 inches thick and distant from one another by 0.0017 inch". He then conjectured that it would be possible to build a replica of it. Volta became aware of the paper, and spent time trying to make a non-metallic battery, using discs of bone soaked with fresh water, sulphuric acid and potash. He had previously verified that a sequence of dissimilar moist conductors could generate an electrical current by contact forces, just as dissimilar metals would.

But success came in 1799 only with what was a mostly metallic construction, consisting of silver and zinc discs separated by brine soaked cloth or paper. The discovery of the repetitive assembly of silver and zinc separated by moist material was almost certainly purely experimental, since there was no theoretical guide which would have suggested it. But this discovery ranks among the highest in the annals of electricity. It provided for the first time a continuous supply of electric current, albeit at a low voltage. But this voltage could be increased easily by adding to the number of silver and zinc discs. A pile with forty or fifty pairs, said Volta, would give anyone who touched its extremities, about the same sensation he could enjoy grasping an electric fish.

This outstanding invention had been obtained in the midst of political turmoil within Lombardy. In 1796 France had invaded Italy under Napoleon, and Volta was chosen as one of a delegation to honour him. Two or three years later Volta was given the Professorship of Experimental Physics at Pavia, about twenty-five miles south of Milan. Then the Austrians returned and closed the University at Pavia. But thirteen months later the French were back and Napoleon reopened the University. Volta proposed in 1800 that he and a colleague should go personally to Paris to express the gratitude of the University to Napoleon. This visit was postponed for a year, during which the publication of Volta's paper on the voltaic pile had taken place. Consequently the visit became a triumph. Napoleon himself attended lectures given by Volta at the Paris Academy, and proposed the award of a gold medal to him. He also gave him a pension and made him a Count and Senator of the Kingdom of Italy, of which Napoleon later became King. There were significant financial advantages in these last honours. While his annual salary at Pavia was about

5000 lire, which was increased by the annual pension of 4000 lire from Napoleon, the Senatorial salary was 24,000 lire.

The European Effect of Volta's Invention

Because Volta was an Anglophile, and possibly because he had received the Copley medal from the Royal Society of London, he decided to publish the discovery of his voltaic cell in England. This was done in two parts, possibly to avoid the loss of a single manuscript in transit, because Britain and France were at war at this time. The first part arrived in London in mid April 1800, and the second part about six weeks later. Nevertheless both parts were written in French. The recipient in London was the President of the Royal Society, Sir Joseph Banks. He, in turn, still in April 1800, passed a copy of the first part of the paper to William Nicholson, who shared it with Anthony Carlisle, a surgeon in London. Both these men then began a series of experiments with a voltaic pile made up initially of seventeen half crowns, and the same number of pieces of zinc, with cardboard soaked in salt water separating the two metals. After several experiments, including the use of water from the New River bringing water from Hertfordshire to the City, they separated out two volumes of hydrogen gas to one volume of oxygen gas from the electrolysis of the water. Thus began the new science of electrochemistry. Their results were published in William Nicholson's own Journal, along with instructions for making a voltaic cell, which Volta had never patented.

Volta's Religious Life

It may be helpful to remember that Volta's life spanned that of Napoleon Bonaparte (1769-1821). It is recorded that Napoleon expressed his creed in the following way: "I believe that man is the product of the fluids of the atmosphere, (including the electrics), that the brain produces these fluids and gives life, and that after death they return to the ether." This cannot be said to be helpfully informative since neither the fluids nor the ether are defined.

It has been previously stated that Volta was educated within a Jesuit environment, and was removed from that when a Dominican uncle foresaw the danger of him being taken permanently within the Jesuit fold. The Popes Paul III and Julius III, seeing what support the Jesuits might provide against the Reformation, had granted them privileges such as no body of men had ever before obtained. But the learned men of the age began to oppose the influence they had acquired, and their intriguing spirit made them objects of suspicion in many countries. Thus in 1759 the Jesuits were expelled from Portugal, and in 1764 they were expelled from all French geographical possessions. Finally in 1773 Pope Clement XIV abolished the Jesuits from all the States of Christendom.

As far as Volta was concerned he was given Thomas à Kempis's book *Imitatio Christi* to read when he was aged sixteen. But he had his own ideas on religion, and at the same age recorded his belief that "animals too had a soul, conceived of as a true spiritual substance". In later life at age twenty-nine he produced a catalogue of books in the Jesuit library in Como where he had been educated. To show its breadth the list included books by Galileo, Descartes, Musschenbroek, Nollet and Gravesande.

Four years later in 1778 Volta obtained his Chair in the University of Pavia, where he became acquainted with the Italian Jansenist movement. This represented an evangelical approach to Christianity within the Catholic Church. At this stage of his life Volta was unmarried, and he did not finally marry until 1794. This was because the girl he wished to marry was of a lower social class, and the view of his family closely connected with the church effectively debarred such a marriage. Volta's view was, "A dejected mind and upset heart are hardly compatible with the maxims of Christianity, while a marriage to this girl, dictated by genuine sentiment could be." In this viewpoint Volta was rebelling against the Catholic Church, as the Jansenists also were. But the Jansenist views in what was the leading educational establishment in Lombardy were also in keeping with those of Joseph II (1741-1790), the German Emperor son of Francis I and Maria Theresa. He began an extensive series of reforms when he became German Emperor in 1765, but only in 1780, in the estates of the House of Austria when he succeeded his Mother. But these reforms were unpopular and he was forced to give many of them up. However, he asserted the right of the State to regulate all religious matters affecting public administration.

Consequently when Austrian power was restored in Lombardy in 1815 after the Battle of Waterloo, and Volta was under some suspicion for the closeness of his connection with Napoleon, he found it necessary to reassert his belief in Catholicism. This he did by asserting that while he believed in the importance of grace for Salvation, he also believed in the role of "good actions", so rejecting the Jansenist idea that divine grace alone can save human beings.

The Battle of Waterloo was also a turning point within France for progress in understanding electricity. For in parallel with Volta's progress in Italy a succession of three scientists in France took over the European leadership within electrostatics. These three were Coulomb, Laplace and Poisson, who will now be considered in turn.

Charles Augustin Coulomb (1736-1806)

C. A. Coulomb was born in Angouleme, about seventy miles North East of Bordeaux. His father had been in the army, but had become a tax-farmer based in Paris. As a result Charles Augustus was able to attend there the College Mazarin, which was founded by the will of Cardinal Mazarin, who had been the virtual ruler of France for many years. Although the Cardinal's tax raising system made him very unpopular, he supported scholarship to the extent that he made his famous library open to literary men, and gave pensions to both the mathematician d'Alembert and the scientist Lavoisier.

It appears that Coulomb also attended mathematical lectures at the College Royal de France because of the reputation there of Pierre Charles le Monnier. But when he announced that he too wished to become a mathematician this caused a family upset. It was his mother's wish that he should study medicine, and he was forced to move to his father's house in Montpelier, about one hundred miles west of Marseilles. The parents' separation had been the result of the father's speculative financial losses.

In Montpelier family connections allowed Charles Augustin to be introduced to the scientific circle there. The Societe Royale des Sciences de Montpelier was proud of its being the second Royal scientific society in France. It had five sections, in mathematics, anatomy, chemistry, botany and physics, and in each section there could be fifteen adjunct members in addition to three regular members. Adjunct members had to be at least twenty years of age. In Coulomb's case he submitted a mathematical memoir entitled "Geometrical essay on mean proportional curves" at age twenty-one and was accepted as an Adjunct Member. During his sixteen months of membership he submitted five memoirs, at least two of which were on observational astronomy, for which the Society was well equipped.

Since Coulomb needed financial support, the options he considered were the Church, in which he could have become an abbe, or becoming a civil or military engineer. In civil engineering the work would have involved the building of bridges and roads. By contrast military engineering would have included the building of forts, shipyards and ports, which would offer a wider range of intellectual challenges. Consequently Coulomb's decision was to enter the Ecole du Genie, for which at the time the most famous school in Europe was located at Mezieres. Entry to this school was by entrance examination, for which the best preparation was available in Paris. This Coulomb undertook, and started his course in Meziere in February 1760, staying there until he graduated in November 1761 with the rank of Lieutenant en Premier. Courses given in the college included carpentry, stonecutting, drafting, physics, mathematics including calculus and hydrodynamics. Coulomb was posted to Brest initially, but because of illness on the part of an engineer in Martinique was transferred there in 1764.

During the Seven Years' War, Port Royal in Martinique was bombarded by the British fleet and effectively destroyed. Following the Treaty of Paris in 1763 the whole island was returned to the French, who decided on a major reconstruction of the fortifications of the town. The scheme for this work was put out to competition and was won by an engineer with thirty years' experience. But his designs would have cost fifteen million livres. Such a cost brought a rethink which was carried out by a panel which included the Governor of the Isle of Martinique, the original designer, Coulomb and other engineers. Of fourteen people on this committee only four supported the revised plan, the remaining ten abstaining. The four supporters included the original designer and Coulomb, who was then put in charge of the fortifications on the mountain behind Port Royal. These fortifications formed the linchpin of the island's defences. It took eight years for the work to be completed, with Coulomb in charge of everything, stonecutting, rock removal, masonry, retaining walls, vaults and the labour force.

For the next two years Coulomb was posted to Cherbourg. This allowed free time in which to write a paper on engineering mechanics, which was presented to the Academy of Sciences in Paris. It was while in Cherbourg also that he wrote "Investigations of the Best Method of Making Magnetic Needles". This was the subject of an Academy prize contest set in 1773 for submission in 1775. Since no prize was awarded then, the subject was reset for submission in 1777, and Coulomb's paper was one of two which shared a joint award. A further achievement during the time spent in Cherbourg was a paper he wrote describing the use to which engineers could be put during peacetime. Coulomb wrote, "The more they are occupied during peacetime, the more they will acquire the means to be useful during wartime.—It is the same with the Spirit as with the Body; both have need of exercise for their preservation."

In 1777 Coulomb was transferred from Cherbourg to Besancon, capital of the department Doubs. Here he was able to write further civil and mechanical engineering memoirs, before being transferred to Rochefort on the west coast of France in 1779. This allowed him to conduct experiments on friction in the shipyard there, which provided material for another Academy prize in 1781. With this double achievement Coulomb felt able to apply for a permanent position in Paris as an engineer at the Bastille. One advantage of this was that he would be relatively free to attend Academy of Sciences meetings, to membership of which he had recently been elected. It also gave access to books and the company of fellow scientists. It was also the beginning of a transfer of his interests from engineering to science.

Coulomb's Work in Electricity and Magnetism

When Coulomb was stationed at Cherbourg he submitted his paper on "Investigations of the Best Method of Making Magnetic Needles" for the Academy of Sciences Prize to be awarded in 1777. It was desired that improved compass needles could be manufactured, and that an explanation would be found for the daily variation of the earth's magnetic field. But before that could be done it was necessary to establish how much that variation was. The existing method of measurement used a pivoted magnetic needle in which the frictional force might vary as much as the variation it was designed to measure. But Coulomb had experience of twisting in ropes, and implemented first a fine silk thread for the suspension of the needle. But this proved too sensitive, and was replaced by a wire. Within certain limits Coulomb found that the torque exerted by a twisted wire was proportional to the fourth power of its radius, and to its angle of torsion, divided by the length of the wire. This enabled Coulomb to obtain results for the repulsive force between like charges, which proved that it varied with the inverse square of the distance between them. It also agreed with results previously obtained by Robison, although this law is now known as Coulomb's law. This full memoir for like charges was written in 1785.

With regard to the force between opposite charges, considered in Coulomb's second memoir (1787), this proved more difficult to measure because the attractive force between the pith balls tended to bring them together for all separation distances. Consequently he used a method of measuring the oscillation frequency of a thin horizontal dielectric rod at one end of which was placed a small metallic disc. The rod was suspended from a vertical silk thread at variable distances from an electrified sphere with an opposite charge to that of the disc. He found that the period of oscillation of the rod was directly proportional to the separation distance between the disc and sphere. This indicated that the force of attraction was proportional to both the inverse square of the period of oscillation and the distance between the bodies. In this second memoir Coulomb also extended his results to the forces between magnetic poles, as he had done previously for electric charges.

In his third memoir (1787) Coulomb dealt with the matter of charge leakage. He had observed that the mechanical force between two charged bodies diminished with time. In so far as this indicated a charge loss this could be attributed to loss through the air or leakage along the surface of the supporting dielectric. From his experiments he found that the rate of charge loss was proportional to the charge, i.e. the charge decreased exponentially with time for a given degree of humidity.

In the first part of his fourth memoir (1787) Coulomb described experiments which showed that electricity spread itself in the same way over identically shaped conducting bodies, irrespective of the particular conducting material. When the material was a good conductor the charge was distributed in an imperceptibly short time, but with a poor conductor like paper the distribution took several seconds. In the second part of the memoir Coulomb showed that on a charged conducting body the electricity distributes itself only on the outside surface of the body and is null within. He was able to show this both experimentally and theoretically.

The aim of the fifth memoir (1788) was to show how the charge densities varied on two conducting spheres in contact after being charged and then separated. Clearly if the spheres are of equal radius the total charges and charge densities on the spheres will be equal. But if, for example, one sphere is of twice the radius of the other, Coulomb's measured results showed that the charge density on the smaller sphere was 8% larger than on the bigger sphere. Similarly if one sphere has a radius eight times the other, his results showed the charge density on the smaller sphere was 65% larger than on the larger. It will be shown in a later section how these results compare with the corresponding theoretical values obtained by Poisson over twenty years later as published in the Institute in 1811. It should be noted here that it is the charge densities which have been measured. Obviously on each sphere its potential will be constant over its full surface.

Coulomb then extended his measurements to give the charge density on each of two spheres which were placed in contact and then charged. He did this for spheres of different radii. When the spheres had equal radii he measured the ratio of the charge density at a point 180 degrees from the point of contact to the mean charge density and found this to be 1.27. Poisson's theoretical result obtained later was 1.3. But when the ratio of sphere diameters was eight to one this ratio increased to 3.18 using the mean charge density for the larger sphere. Poisson's theoretical result in this case was 3.1.

In the sixth memoir (1790) Coulomb measured the ratio of the charge densities in the middle of a long line of twenty-four contacting spheres to their value at the end of the line. He found the charge density at the end to rise, to a value of 1.75 that at the middle. This result is to be expected to make up for the absence of those spheres beyond the end of the line. He further considered a contacting junction between spheres and cylinders. Thereafter he extended the study to consider the case of an imagined cloud with a radius of 1000 feet in contact with a dielectric string with a thickness of 1/6 inch. His calculations showed that the mean density on the surface of the string was 27,000 times that of the cloud. Since he had already found that the ratio of the density at the ends of a cylinder to that of its mean was 2.3:1 this gives a density at the end of the string as 62,000 times that in the cloud. The implication with regard to flying a kite in a thunderstorm is obvious. It should not be done.

Following the Revolution in 1789 the Academy des Sciences was abolished in 1793. It was later reformed as the Institut de France, and Coulomb was elected to its membership in 1795. Thereafter he contributed to the reform of Education in France, as Inspector General of Public Instruction. In 1802 Napoleon ordered a plan for the reorganisation of French education.

There would be four levels, Ecoles Primaires and Secondaires, organised by the towns, and Lycees and Ecoles Speciales organised by the state. A six-man commission was appointed to enforce this plan, and Coulomb was a member of this from its inception until he died in 1806.

Of the four levels the Lycees were the most privileged group. Chaplains were appointed by Napoleon for each one. The rigorous religious training was intended not so much for the salvation of the students as for their control. The course lasted three years, and the best students were given two additional years for further study. Initially forty-five Lycees were proposed for France and Belgium, although only twenty-nine were fully operational by 1806. There was to be a chapel in each Lycee which was to accommodate one hundred and fifty pupils and fifteen hundred books. The commission were to recommend two candidates for each staff position, with Napoleon making the final choice.

Coulomb may be said to have been responsible, by example, for the birth of modern physics in France.

Pierre Simon Laplace (1749-1827)

Pierre Simon Laplace lost his mother at an early, but unknown, age, and had an unrewarding relationship with his father and siblings. So much so that none of them were present at their father's funeral.

He was born at Beaumont-en-Auge in Normandy, where his family had farmed for generations. The land was fertile, but in 1774, when his father remarried, he took on the local inn in Beaumont, where he also owned a local farmhouse. From the absence of references in Pierre Simon's later life to his youth it is likely that he suffered emotionally in these early years. His father became Mayor of Beaumont for over thirty years, and it may be that he had little time to spend with his son. Nevertheless he did wish his son to have a good education. And to achieve this he made use of his brother, Pierre's uncle, Louis Laplace, who was an ordained Abbe, and who assisted at the local school run by the Benedictine monks. Because Pierre was a local boy he benefited from the patrimony of the Duc of Orleans. This allowed him to have free education at the local school. Associated with this there was a requirement that the free scholars would pray daily for the Duc, and compliment him with a formal address on every public occasion.

During his school life Pierre Simon lost both his younger brother and his uncle Louise. Possibly as a result of this he developed a rather austere attitude to life. This attitude was accentuated by the working rules of the Benedictine school. Their work ethic is encapsulated in the phrase "working like a Benedictine". In school the teaching day lasted for eight hours. Silence was obligatory during meal times, and the school recommended to parents that the six-week summer vacation be skipped every second year.

Pierre Simon was sent to the University of Caen when he was around age sixteen, with the expectation that he would become a priest. A story is told about a return visit to Caen in 1813 when he was sixty-four years of age. Seeing a boy doing penance for a minor misdemeanour, he addressed the priest and asked for forgiveness for the boy in the name of his own service to the church so many years before.

As a student at Caen he studied for four years until 1769 when he took his Master's degree. During this period he observed the dissensions in religious views between members of staff. The traditionalists were opposed by younger staff who had acquired knowledge of the mathematical writings of Newton. This may have deterred Laplace from following a career in the Church which he could have done. Entry to the Church at that time was, at least sometimes, the result of calculated self-interest, without considering the spiritual mission of the priesthood. He himself liked mathematics and he took the plunge to be adventurous. Armed with a letter of introduction from his teachers at Caen he decided to visit in Paris the grand old man of mathematics, d'Alembert.

Laplace was twenty years of age when he first met d'Alembert, who was then aged fifty-two. D'Alembert had written a paper on the inertia of a body moving through a vacuum. After studying it Laplace generalised it to include the case of a body moving through a resisting medium. This impressed d'Alembert sufficiently to arrange an appointment for Laplace, whom he referred to as the Abbe Laplace, as a mathematics teacher at the Ecole Militaire. He was asked to wear clerical garb, including his ecclesiastical collar. The Library in 1776 had over four thousand volumes, and included the transactions of the major learned Societies at Paris, London, Berlin, St Petersberg, Turin and Gottingen. Moreover, the authors of the textbooks used for the students were all members of the Academie des Sciences. In all there were fourteen instructors in mathematics, of whom only one was research oriented.

But in the School Library he was able to study three books on the Calculus by Leonard Euler, viz. Introduction to Infinitesimal Calculus, Foundations of Differential Calculus, and Foundations of Integral Calculus. Euler was envied by d'Alembert. Further, the French mathematicians Lagrange and Legendre were also brought to his notice.

It became clear to Laplace that he should strive to become a Member of the Academie des Sciences. To achieve this he submitted over a period of three years no fewer than thirteen research papers before election came his way. These papers covered mathematics, celestial mechanics and probability theory.

Immediately prior to the French Revolution of 1789 Laplace had married into a family which had been recently ennobled, and who were naturally concerned about their material assets. Laplace was naturally a very cautious man, and even in his career had assessed his likely career prospects before leaving the Church. But the Revolution was cataclysmic. It even had a major effect on the Academie des Sciences which was required to draft a new Constitution for itself. Laplace was appointed to this Committee. As might have been expected, dissensions arose and the Academy was actually closed down in 1793. This was the year also when the King, Louis XVI, was guillotined, as was also the Queen, and Christianity was formally abolished. So began the Reign of Terror which lasted until 1794. Among its additional victims was the French Lavoisier, who was the first person to establish the terminology of Chemistry. During his lifetime he had been appointed Farmer General of Taxes in 1769, and although in 1790 he had sat on the Commission of Weights and Measures, he was accused as an ex-Farmer General and guillotined.

With regard to teaching, in the new environment it was clear that a substitute had to be found for teaching young men who had previously been taught by the religious establishment. One outcome was the creation of the Ecole Centrale des Travaux Publiques, which later became the famous Ecole Polytechnique. But they also conceived the idea of the Ecole Normale which in the short term would provide teachers who would supply a sound basic education. This proposal was enacted in 1794. Laplace, along with Monge and Lagrange, was asked to provide a framework for the teaching of all elementary mathematics in France. The Republic was making known its acceptance of the view that France needed scientists.

Laplace gave the inaugural lecture in January 1795 to about seven hundred mature students. Lectures in other disciplines were also provided. All the sessions were recorded in shorthand, and published afterwards so as to be available nationally. In Laplace's own lectures he used no equations, but chose what has been called "a literary rendition of an abstract art". This entailed using memoirs of e.g. Descartes, Newton and Euler.

In addition Laplace provided a popular treatise entitled *Account of the System of the World*. This was translated into German in 1797, into English in 1809, and later into Russian and Chinese. It has been described by Hahn in the following way: "He takes the reader by the hand from the appearance of motion of the heavenly bodies to their real motions; from those to the abstract laws of physics; and from those to the law of gravitation, no longer presented as a conjecture, but now taken as a veritable cause of motion, a grand principle of nature."

The combination of the publicity associated with the Ecole Normale and the Account of the System of the World meant that Laplace became a very public figure in 1795. Because of this his influence in Government circles increased, and he played a major part in the formation of the Institut National des Sciences et des Arts. This restored the intellectual side of life in France, and in its first Annual Report given in 1796 it is interesting to observe the careful political note struck by Laplace, as translated by Hahn:

"This account would be incomplete if we did not refer to our efforts to propagate the eternal principles of justice and equality that are at the root of the French Constitution—We declare that it has no more sincere champions than savants and artists. Nature, which is the object of their constant concern, reveals at every turn the right and dignity of man—men are passionately stimulated by all that is grand and orderly, so that equally far from servitude and anarchy, everything takes them to a Government that steers a middle course between these two extremes, and whose existence is intimately linked to the progress of science and the arts, without which there is no desirable liberty, no happiness."

Another of Laplace's publicity seeking activities was the calling of an international conference to certify the validity of the metric system. In this proposal he was backed by Napoleon who had been a student of Laplace. In 1785 Laplace had recommended that Napoleon Bonaparte be given a commission in the artillery. Further he had been elected as a member of the Institut National des Sciences et des Arts. And after his successful invasion of Egypt he had set up a research institute there which was modelled on its French counterpart.

After Napoleon's return from Egypt he became first Consul in France in November 1799. Within three days Laplace had been appointed Minister of the Interior, but retained his post for only six weeks. It is believed that he over concerned himself with details, rather than the broader picture. However, he was appointed in the same month to the Senate, of which he became successively Secretary, Vice President, President and finally Chancellor in 1803. For this last post the annual salary was seventy-two thousand francs.

After Napoleon's defeat at Leipzig in October 1813 the Allied forces entered Paris in March 1814 and the Senate dethroned Napoleon and exiled him to Elba. But before this happened Napoleon's relationship with Laplace had suffered severely. After the defeat at Leipzig Napoleon remarked to Laplace:

"Oh, I see you have grown thin."

"Sire," Laplace replied, "I have lost my daughter in childbirth."

"Oh," said Napoleon, "that is not a reason for losing weight. You are a mathematician. Put this event in an equation, and you will find it adds up to zero."

This insensitivity on Napoleon's part hurt Laplace.

After Napoleon was sent into exile Laplace was deputed by the Senate to welcome King Louis XVIII on the return of the monarchy. Although Napoleon escaped from Elba in March 1815 he was defeated at Waterloo in June of that year.

In April 1816 Laplace was elected to the Academie Francaise partly because of the success of his Account of the System of the World. This brought him into contact with Mary Somerville (1780-1872) the Scottish feminist and writer on the physical sciences, who was the daughter of Admiral Fairfax. Laplace and his wife entertained her at their house in 1817, though it was not until 1827 that she produced her first book *Mechanism of the Heavens*. Laplace's wife was twenty years younger than him and was a practising Catholic.

Towards the end of his life Laplace enquired about Isaac Newton's views on religion. This was a subject on which Newton had written more words than he had on physics. His views were what the Christian Church describes as Socinian, i.e. they postulated that Jesus Christ had been created by the Father, rather than having always been co-eternal with the Father. Nevertheless Newton belonged to one branch of the Christian church, and this puzzled Laplace, who wondered if Newton's mental powers had been adversely affected. But one of Laplace's protégés was Jean Theodore Maurice, a Genevan astronomer with an evangelical background. He had drawn to Laplace's notice the work of Thomas Chalmers' *The Evidence and Authority of the Christian Revelation*, which had been translated into French. This had not convinced Laplace, nor had Gilbert Burnet's *A Rational Method for Proving the Truth of the Christian Religion*. What held Laplace back was his unshakeable belief in the total completeness of his own mathematical views, which did not allow the existence of the supernatural. But there is also evidence that Laplace accepted the beauty and value of Christianity as a civilising influence, and that he would have been glad to share "the happy conviction that God had given to Maurice".

Aphorisms of Laplace

(1) What we know is not much. What we do not know is immense.
(2) Man follows only phantoms.
(3) Nature laughs at the difficulties of integration.
(4) Read Euler. He is our master in everything.
(5) Such is the advantage of a well-constructed language that its simplified notation often becomes the source of profound theories.
(6) Napoleon: You have written this huge book on the system of the world without once mentioning the author of the universe.
    Laplace: Sire, I had no need of that hypothesis.
    Lagrange: Ah, but that is a fine hypothesis. It explains so many things.
(7) Napier's logarithms by shortening the labours doubled the life of the astronomer.
(8) It is India that gave us the ingenious method of expressing all numbers by means of ten symbols, each symbol receiving a value of

position as well as an absolute value; a profound and important idea which appears so simple to us now that we ignore its true merit. But its very simplicity and the great ease which it has lent to computations put our arithmetic in the first rank of useful inventions, and we shall appreciate the grandeur of the achievements the more when we remember that it escaped the genius of Archimedes and Apollonius, two of the greatest men produced by antiquity.

(9)   The theory of probabilities is at bottom nothing but common sense reduced to calculus; it enables us to appreciate with exactness that which accurate minds feel with a sort of instinct for which ofttimes they are unable to account.

(10)   All the effects of Nature are only the mathematical consequences of a small number of immutable laws.

Simeon-Denis Poisson (1781-1840)

Simon-Denis Poisson's parents were not from the nobility. His father had been a soldier and had suffered discrimination by the nobility in the army. After retiring from the army he had bought a lowly paid administrative job in the village of Pithiviers, forty miles south of Paris. He is believed to have taught his son Simeon-Denis to read and write. When the revolution of 1789 broke out his father took over the governing of the village.

The parents wished him to study medicine and Simeon was apprenticed for a time to an uncle in Fontainebleau nearby. In his early training he had to prick veins in cabbage leaves with a lancet. This was followed by pricking blisters. But early on in this experience the patient died within a few hours, and Simon decided to give up medicine. He then enrolled at age fifteen at a central school in Fontainebleau. This brought to light his ability in Mathematics, so that he was encouraged to apply for entry to the Ecole Polytechnique in Paris. In this competitive entry he was placed first.

Poisson's years at the Ecole Polytechnique

During his two years study at the Ecole Polytechnique two incidents are worthy of record. The first is that at an oral examination conducted by

Lagrange, Poisson was asked to prove Newton's Binomial Theorem. Lagrange was one of the two masters of the following generation of French mathematicians, and expected a standard proof. But Poisson insisted on another. Lagrange listened, found Poisson's proof better than his own, adopted it and henceforth used it, always acknowledging its inventor. The second incident is that in less than two years at the Ecole Polytechnique Poisson published a paper on the theory of equations. This enabled him to be given, immediately after his graduation in 1800, the post of assistant master. Two years later he became Deputy Professor and in 1806 Titular Professor. The restoration of Louis XVIII in 1814 caused no problems, because basically Poisson was a scientist and not a politician. When the Revolution of 1830 occurred Poisson once again supported the new government. In this respect Poisson was following the example of Lagrange who had formulated the following rule of conduct: "I believe that, in general, one of the first principles of every wise man is to conform strictly to the rules of the country in which he is living, even when they are unreasonable." This philosophy was put to a severe test when it was decreed in 1793 that all foreigners born in enemy territory, and this included Lagrange, were to be arrested and their property confiscated. Fortunately intervention was made on Laguerre's behalf and he was excused from its operation.

Poisson in Old Age

As an indication of the quality of the man whom Poisson became, the following comment by Cournot who replaced him in 1839 as Chairman of the Jury d'Agregation in Mathematics is significant: "At the very end of his life when it had become painful for him to speak, I saw him almost weep from the chagrin that he had experienced as Chairman of a competitive examination for the Aggregation, for he had become convinced that our young teachers were concerned solely with obtaining a post, and possessed no love for science at all." This suggests that Poisson, an unbeliever in formal religion, had his own ideal of what one aspect of religious behaviour should aspire to.

Poisson's Extension of Laplace's Equation

From 1814-1827 Poisson worked closely with Laplace. In particular he generalised the equation which Laplace had obtained for calculating the attraction exerted by a mass on a unit test mass at a point external to the mass. To do this Laplace had divided the mass by the distance to the external point. This was the potential V at this point, and its derivative gave the attractive force. Poisson's contribution was to calculate the attraction when the point was either inside the mass or on its surface. This result he then transferred to the corresponding electrical situation, when the mass is replaced by the electric charge. Dividing this by the distance to the external observation point as before, gives the electrical potential V at that point. Then the components of force on a unit test charge are given by the slope of the curve of V against distance.

Poisson's Mathematical Memorial in Electrostatics

Poisson's investigations of Electrostatics from this point onwards have been described as a splendid memorial of his genius. To illustrate this he was asked in 1811 to solve the problem of finding the electric charge distribution on the surface of conductors, both when they were in isolation, and also when they were in the presence of each other. Initially he confined himself to the case of these conductors being spheres. This was probably because experimental results for spheres had been obtained by Coulomb between 1785 and 1789. These results included two cases which Poisson believed could be solved theoretically.

The first of these was to find the electrical charge density on each of two spheres which were placed in contact and then charged. It is clear that if the two spheres are of equal radius the charge distribution on each will be identical. But it will not be uniform round the surface as when each sphere is in isolation. For at the point of contact any charge which existed there on one sphere would repel the equal charge on the other sphere. Consequently the charge density at this point of contact must be zero. The density will then rise towards the diametrically opposite point one hundred and eighty degrees away, in such a way that the electrical potential round the surfaces of both spheres will be constant. This is as far as physics alone will go, and where the mathematics has to take over. Poisson's analysis, obtained as a series of Legendre functions, gives the ration of the 180 degree density to the average density round the sphere to be approximately 1.3.

When the radius of one sphere is increased above that of the other, the effect is to increase the electric charge density on the smaller sphere, but leave the density on the larger one more uniform over most of its surface. Thus for a ratio of radii of four, Poisson found that the previous value of 1.3 was increased to 2.5.

The second of the two cases that Poisson dealt with theoretically was calculating the charge density on two charged conducting spheres placed at any distance from each other. An approximate result can be obtained when the spheres are a large distance apart. Thus if the electric field of one sphere is approximately constant and parallel to a diameter of the second sphere, the charge distribution on this second sphere will be approximately cosinusoidal, taking an origin at the centre of the second sphere. But in the general case the solution is not of a simple form. It has nevertheless been quantified by Poisson.

Additional Poisson Contributions

A Poisson distribution is the probability of the number of occurrences of a rare event, but which has many opportunities to occur. When such events occur at random intervals, a knowledge of the mean number of occurrences in unit time, enables the probability of there being any stated number of occurrences in any given interval to be calculated. An electrical example is Shot noise in diodes, and a more recent example is stated to be the number of mutations in a given stretch of DNA after a certain amount of radiation.

CHAPTER FIVE

THOMAS YOUNG : THE QUAKER INFLUENCE

(c.1800 – c.1826)

CHAPTER FOUR ENDED with the mathematical work of Poisson as applied to electrostatic problems. His solutions for the charge distributions on conducting bodies were published in 1824. But some years before this several contrasting experimental and theoretical results were obtained which swung physics in new directions.

The first came in 1800 from Thomas Young (1773-1829), the son of a banker from the village of Milverton in Somerset. He trained as a physician, and advanced the wave theory of light in opposition to the corpuscular theory of Newton. At the time, of course, it was not realised that the mathematical theory of light was identical with the mathematical theory of electricity. This was a discovery to be made later by James Clerk Maxwell.

The second experimental result was obtained, also in the same year 1800, by William Nicholson (1753-1815), a publisher, and Anthony Carlyle (1768-1840), a surgeon. In repeating and extending Volta's experiments on batteries, they found that when a tube of water was introduced into the external circuit formed by the zinc and copper electrodes, that the wire connected to the zinc electrode became oxidised. Also, bubbles of hydrogen were liberated at the other wire. In this way the decomposition of water was illustrated.

This discovery influenced Humphry Davy (1778-1829), the son of a wood-carver in Penzance who became Professor of Chemistry at the Royal Institution in 1801. Davy invented an electromagnet in 1820 and showed in 1821 that the conducting power of a wire was proportional to its cross sectional area, but independent of the shape of its cross-section. This demonstrated that the current passed through the cross-sectional area and not along its surface. He also showed that the conducting power was inversely proportional to its length.

But all Davy's results followed the third experimental discovery of Professor Hans Christian Oersted (1777-1851) in Denmark in 1820. His father was a pharmacist on the island of Langeland in a town with a population of about one thousand. Oersted's discovery was that the flow of electric current in a wire exerted a mechanical force on an adjacent compass needle. This effect of electricity on magnetism was one which he had believed for many years must exist, but initially he had thought that the effect would be associated with the electric voltage rather than the electric current.

Oersted's discovery created an explosion of interest over Europe. In France with its proven mathematical strengths, the response to Oersted's discovery took the form of both theory and additional experiments. Here the fourth important discovery must be credited to Andre Marie Ampere (1775-1836), who held the position of Professor of Mathematics at the Ecole Polytechnique. On the 11th September 1820 he had seen demonstrated in Paris by Arago the result obtained by Oersted in Copenhagen. Within one week he had repeated the experiment, and extended it to the extent that he showed like currents attracted each other. This was the opposite effect to like charges, which repel each other. Moreover he showed that the force of attraction between like currents was proportional to the strength of these currents, and inversely proportional to the square of the distance between them.

In parallel with the work of Ampere in Paris there was a related analysis carried out by a scientific competitor in the person of Jean-Baptiste Biot (1774-1862). He had become Professor of Physics at the College de France through the influence of Laplace, for whom he had undertaken the work of proof reading Laplace's *Mecanique Celeste*. Experimentally he had taken part in 1804 in the first balloon ascent undertaken for scientific purposes. This showed that the earth's magnetic field does not vary significantly with altitude up to the maximum height they achieved of 13,000 feet. In Biot's electrical researches he worked in collaboration with Felix Savart to the extent that the result they obtained for the strength of magnetic field produced by a current element Idl is now known as the Biot-Savart law. This constitutes the fifth important discovery. It says that the maximum strength of this magnetic field due to the current Idl is proportional to the magnitude I of the current and inversely proportional to the square of the distance to the point of observation.

The final component in this group of six seminal experiments covering the period 1800-1840, relates to Augustin Jean Fresnel (1788-1827). He began his experiments on light in 1814, unaware of the earlier work which had been carried out by Thomas Young in Cambridge between 1797 and 1799. This was not surprising both since he was working well away from Paris and he did not speak English. Nevertheless he combined analysis with experiment. In particular in 1821 he formulated a wave equation. Then by making his quantities vary harmonically with time his equations were reduced to a manageable form.

Each of these contributions to the greater understanding of electricity will now be considered in turn.

Thomas Young (1773-1829)

Thomas Young's father was both a cloth merchant and a banker in the village of Milverton in Somerset. His marriage produced ten children, of whom Thomas was the first. It appears that Thomas was brought up mainly by his grandparents in Minehead. He was a child prodigy, having learned to read at age two. By the time he was six, and possibly only four, he had read through the Bible twice. He was gifted with a formidable memory which must inevitably aid progress educationally.

Thomas Young's Schooling

When it came to schooling he attended a number of different schools. Two were boarding schools, the first of which he described as miserable. After spending a year there he stayed at home for the next six months. But during this period he benefited from a neighbour who was a land surveyor, and who allowed him the use of his library. He then moved to a boarding school at Compton in Dorset at age nine and stayed there for four years. During this period "he was in the habit of rising an hour sooner than my school fellows in summer, and of going to bed an hour or two later in winter for the purpose of mastering my lessons for the day". This was an example of the importance of industry from his Quaker background. It was said of him that he did not spend an idle day in the whole of his life. The usher of this school at

Compton was Josiah Jeffrey who lent him Benjamin Martin's *Lectures on Natural Philosophy* and also gave him the use of a lathe. This provided experience in making telescopes and microscopes. From the book he also learned bookbinding and the principles of drawing. During his school career up to the age of thirteen, he was able to study Latin, Greek, French and Italian. But he had also got through six chapters of the Hebrew Bible. Also independently he began to study natural history. In the school holidays he was able to procure a lathe for turning optical glass. When he returned home at age thirteen he started to learn Arabic and Persian. A neighbour also gave him the Lord's Prayer in more than one hundred languages, which pleased him greatly.

His Independent Study and the Influence of his Religious Environment

When he returned home at age thirteen he continued with Hebrew and added to it Chaldean, Syriac, Samaritan, Persian, Turkish and Ethiopic. At this time he benefited socially from the religion of his parents, who were Quakers. The Quaker movement was started by George Fox in 1647. It believes in a measure of God's spirit being given to all men and women, resulting in the sacredness of every person. Their primary religious authority is this Inner Light, and the Bible is secondary in authority to it. The celebration of the Lord's Supper and Baptism are not practised. In the lifetime of Young's parents, Quakers also abstained from alcohol and would not visit places of entertainment.

Because of the importance of this Inner Light to members there was no place in their lives for established authority such as church or state. This meant that Quakerism tends to produce people who are very conscious of their own power over their own lives. But within their own group, they will support each other as do people in other church denominations. This benefited Thomas Young from age 14 to 19, when he was able to spend two thirds of each year in a Quaker country house in Hertfordshire, and the other third in central London during the winter. The Quaker owner of these houses was a David Barclay associated with banking, who wanted a companion to share the education of his twelve-year-old grandson Hudson Gurney. The friendship that developed between the boys lasted all their lives.

David Barclay of Hertfordshire, the Descendant of David Barclay the Laird of Urie

The Quaker connection of David Barclay of Hertfordshire goes back to his ancestor David Barclay (1610-1686), the Laird of Urie in Kincardineshire. The Laird had been brought up as a Presbyterian. He had then fought in the Thirty Years War (1618-1648) under Gustavus Adolphus who supported the Protestant cause for two years before his death in 1632. Later David Barclay supported the Parliamentarians successfully at the battle of Marston Moor (near York) in 1644. But it was twenty two years later in 1666 that he heard George Fox (1624-1691), the founder of the Quaker Movement, preach.

George Fox, at the age of nineteen, believed he had received a divine command to forsake everything and devote himself completely to religion. His religious life was peripatetic, and opposition to him was such that he spent several years in prison. He was a mystic but in a practical way, for which he won much esteem. It was said that Fox had an unfailing perception of character. Even when he could not see people with his eyes, he realised what sort of people were near him. And while he travelled mostly in England, on the Continent and the West Indies, a visit to Scotland produced the following saying: "When first I set my horse's foot atop of Scottish ground, I felt the seed of God sparkle about me like innumerable sparks of fire."

Certainly David Barclay in 1666 was converted through him, although there may also have been a contributory factor later, when, because of his conversion to Quakerism, he was imprisoned, and was influenced by another prisoner. One result of the conversion has been brought out in John Greenleaf Whittier's poem "Barclay of Urie." As a Quaker he became the object of persecution and abuse at the hands of the magistrates and the populace. None bore the indignities of the mob with greater patience and nobleness of soul than this once proud gentleman and soldier. One of his friends, on an occasion of uncommon rudeness, lamented that he should be treated so harshly in his old age who had been so honoured before. "I find more satisfaction," said Barclay, "as well as honour in being thus insulted for my religious principles, than when, a few years ago, it was usual for the magistrates, as I passed the city of Aberdeen, to meet me on the road and conduct me to public entertainment in their hall, and then escort me out again, to gain my favour." Seven verses taken from Whittier's poem follow:

Up the streets of Aberdeen,
By the kirk and college green,
Rode the Laird of Urie;
Close behind him, close beside,
Foul of mouth and evil-eyed,
Pressed the mob in fury.

Flouted him the drunken churl,
Jeered at him the serving girl,
Prompt to please her master;
And the begging carlin, late
Fed and clothed at Urie's gate,
Cursed him as he passed her

But from out the thickening crowd
Cried a sudden voice and loud:
"Barclay! Ho! a Barclay!"
And the old man at his side
Saw a comrade, battle tried,
Scarred and sunburned darkly;

Who with ready weapon bare,
Fronting to the troopers there
Cried aloud "God save us,
Call ye coward him who stood
Ankle deep in Lutzen's blood,
With the brave Gustavus?"

"Nay, I do not need thy sword,
Comrade mine", said Urie's lord;
"Put it up I pray thee:
Passive to His holy will,
Trust I in my Master still,
Even though he slay me.

So the Laird of Urie said,
Turning slow his horse's head
Towards the Tolbooth prison,

> Where through iron gates, he heard
> Poor disciples of the Word
> Preach of Christ arisen!

Since Thomas Young resided at David Barclay's home either at Youngsbury, or in London, for five years from age 14 to 19, it was inevitable that the Quaker influence would have some effect on his life. While he later joined the Anglican Church, this may well have been a reaction against the practical rather than the spiritual side of the Quaker beliefs.

Thomas Young's Studies at Youngsbury

A classics tutor was employed for the boys, and from him Young developed a beautiful Greek script which was to prove useful in his later work of deciphering the Rosetta Stone. In his mathematical studies Young read both Newton's *Principia* and his *Optics* at age 17. He had a firm belief in the superiority of self-education to that of study in class with a teacher. His later comment in a letter to his brother confirms this. "Masters and mistresses are very necessary to compensate for want of inclination and exertion: but whoever would arrive at excellence must be self-taught." This is reminiscent of the words of Sylvanus Thompson with regard to learning the calculus: "What any fool can do, any other fool can do."

When he was aged fifteen to sixteen, however, a medical cloud developed in Thomas Young's life. He developed pulmonary tuberculosis, a disease which was so commonly fatal that it deterred some doctors from even attempting a cure. He was kept for two years on a diet of milk, buttermilk, eggs and vegetables with a very little weak broth. (Little more than "water in disguise", he wrote later.) But he had the benefit of being treated by two experienced doctors. One was the doctor who had inoculated Empress Catherine II of Russia against smallpox. The other was his great-uncle Dr Richard Brocklesby, who included among his patients the lexicographer Samuel Johnson and Edmund Burke the political writer and statesman. Richard Brocklesby disapproved of Thomas Young's abstaining from the use of sugar on account of its being obtained from the work of slaves in the West Indies. From the age of under fourteen Thomas Young was an advocate of the abolition of slavery, and maintained his abstinence from sugar for seven years. Around that time Hudson Gurney's grandfather liberated thirty slaves from a property he had inherited, and Thomas Young then started to use sugar again.

Richard Brocklesby's circle of friends in London enabled Thomas Young to meet among others, Edmund Burke and the painter Sir Joshua Reynolds. Edmund Burke asked him to write in Greek, King Lear's imprecations on his daughter Goneril:

> "Hear, nature, hear, dear goddess, hear!
> Suspend thy purpose, if thou didst intend
> To make this creature fruitful!
> Into her womb convey sterility!
> Dry up in her the organs of increase;
> And from her derogate body never spring
> A babe to honour her! If she must teem,
> Create her child of spleen, that it may live
> And be a thwart disnatured torment to her!
> Let it stamp wrinkles in her brow of youth;
> With cadent tears fret channels in her cheeks;
> Turn all her mother's pains and benefits
> To laughter and contempt—that she may feel
> How sharper than a serpent's tooth it is
> To have a thankless child!"

On the occasion of one dinner at Dr Brochelsby's house in London one of the guests read out Dr Johnson's Latin poem written on completion of his great dictionary. No doubt Thomas Young appreciated it, and it might have been expected, in view of his prowess in the classics, that he would have opted to continue their study at University. But instead he chose medicine, which was considered at the time to require a classical training. For this his great-uncle, who had no children of his own, had undertaken to pay the fees. And his own father did not object, though conscious of his own responsibilities as a parent, in the matter.

Thomas Young in London—Anatomy, Midwifery and Botany

In 1792, when Thomas Young started his medical training, there were no medical schools associated with the hospitals in London, and regular medical lectures were offered only in St Bartholomew's Hospital. But there were private lecture courses, and perhaps the most famous of these was the Hunterian School of Anatomy. This was founded in 1746 by William Hunter, born in East Kilbride in 1718. He matriculated at the University of Glasgow and began his medical studies in Hamilton around 1736. He also attended medical lectures in Edinburgh before going to London in 1740. In 1748 he became surgeon-man-midwife to the Middlesex hospital, where one of his patients in 1762 was Queen Charlotte, wife of King George III.

William Hunter's teaching has been said to have revivified the teaching of anatomy in Britain. One of his innovations was the provision of a cadaver to each student, a practice which he had observed in Paris. This provision was not easily maintained when the body lasted little more than one week before it decomposed. The source of such cadavers included public hangings and cemeteries, but since 1783 public hangings had been abolished. This resulted in body snatching which was finally controlled through the passing of the Anatomy Act of 1832.

William had a brother John who specialised in surgery rather than obstetrics. His fame extended to his being buried in Westminster Abbey. But since John had ceased to lecture in 1778, the School had thereafter been run by former pupils of William. The staff then used notes written by John, who lived until 1793, the year after Thomas Young entered the School.

In addition to attending the Hunterian School lectures Thomas Young also enrolled for lectures in midwifery and botany in St Bartholomew's Hospital. While there he obtained his first ward experience, going round with the physicians. He also wrote his first major paper, which was read to the Royal Society in 1793 by Dr Brocklesby, when Thomas was only nineteen years of age. The subject of the paper was "Observations on Vision" and related to the accommodation performed by the eye in focusing on objects at different distances. In the following year Thomas Young was elected a Fellow of the Royal Society.

Thomas Young in Edinburgh (1794-1795)

The possession of a university degree was becoming important in order to distinguish medical people from quacks and charlatans. Dr Brocklesby had studied at both Edinburgh and Leiden, so that it was not surprising that Thomas would follow him to Edinburgh. At that time Oxford and Cambridge Universities did not have medical schools, and in any case would not grant degrees to Quakers on doctrinal grounds.

The Medical School in Edinburgh had been founded in 1726, and 17,000 medical students had studied there in its first century. The lectures were in English, unlike those in Oxford or Cambridge, which were in Latin. It has been said that the strength of an Edinburgh training lay in its imparting the elements of anatomy, surgery, chemistry, medical theory and practice. The three years training available was good for turning out general practitioners. But Young did not intend to become such. One year only was his intention, and in that year his closest relationship was with the Professor of Greek. This produced a second edition of the Professor's anthology of Greek poetry. Young also took up theatre going, dancing and the flute, which displeased his Quaker connections.

At the end of the Medical lectures Young went on a two-month journey on horseback as far north as John O'Groats and as far west as the island of Mull. To do this he took advantage of the hospitality that was provided for him through his social connections

For the next year of study he went to Gottingen. The University there was founded in 1737 by King George II of Britain who was also Elector of Hanover. It had a substantial medical school, and its library was the second largest in Europe. This reputation for scholarship has been maintained over the years. It has produced more Nobel Prize winners than any other city in Europe except for Stockholm—not because the Nobel Prize winners were necessarily undergraduates there, but they had worked in Gottingen at some period in their professional lives. Even in the field of theology it was the first University to offer Karl Barth a chair after he had published his book on *Romans*. A tablet commemorating Thomas Young has been placed in the University in Gottingen.

By April 1796 Thomas Young had submitted to Gottingen University a successful dissertation on the organs of the human voice. He submitted an alphabet of forty-seven letters, which either alone or in combination could produce the total range of sounds that the voice could pronounce. This, he submitted, would be useful for writing down the unwritten tribal languages of Africa or America. It is said that the examination of the dissertation lasted for four or five hours, but that the table round which they all sat was "well furnished with cakes, sweetmeats and wine". At the ensuing graduation, since the normal oath was waived because of his Quaker religion, he was "declared to be married to Hygieia the Goddess of Health, and created doctor of physic, surgery and midwifery".

Thomas Young in Cambridge (1797-1799)

During his four years training in London, Edinburgh and Gottingen, the regulations for becoming a Licentiate of the College of Physics had changed, so that attendance at a single University for two years was now required. It appears that Thomas Young chose Cambridge to obtain his M.B. degree in order to facilitate the procurement of his F.R.C.P. in later years, plus a hoped for M.D. from either Oxford or Cambridge. Fellowship of the Royal College of Physicians tended to be in the hands of Oxford or Cambridge men.

But choosing to go to Cambridge involved crossing a religious hurdle. This was because every candidate for a bachelor degree at Cambridge had to declare that he was a bona fide member of the Church of England. The consequence of this was that many members of dissenting churches chose to go into industry rather than university. Examples of this are the industrial firms of Fry and Cadbury, whose families chose also to avoid products associated with alcohol. But Thomas Young probably recognised that the spiritual side of the Church of England was the same as that of the Quakers, and taking the oath allowed him freedom from the restrictive practices which the Quakers imposed on themselves.

The teaching offered in medical subjects in Cambridge was of a lower standard than was available in either London or Edinburgh, though it was here that blood pressure was first measured, and Cambridge was a pioneer in blood transfusion. When Thomas Young entered the University he was admitted as a Fellow Commoner. This gave him the privilege of sitting at the same table as the Fellows, and thus benefiting from their conversation. More generally his other social activities included the Emmanuel College Parlour, which was open to those who loved pipes, tobacco and cheerful conversation.

Indeed, within six months of arriving in Cambridge he was elected President of this Parlour. Their activities also included the placing of bets. One was "that Young does not draw with a pen one hundred lines in the space of an inch". It was said that he rarely associated with the young men of the college, who, perhaps not surprisingly, called him "Phenomenon Young." It is said that he made mediocrities uneasy, and to cover their unease they belittled and ridiculed him.

With regard to academic work Thomas Young did not think it necessary to attend lectures relating to subjects which he had previously studied. It will be recalled that he was already a Fellow of the Royal Society and had a lecture presented there on his behalf on the subject of vision, and the accommodation of the eye. It appears that his study time in Cambridge was much spent in solitary reading and doing experiments in physics. Certainly initially this would have dealt with sound, the subject of his dissertation at Gottingen, and was thereafter generalised to fluids. He was also disappointed in the level of mathematics at Cambridge, knowing that Continental mathematicians greatly surpassed the English. But the seminal studies he began in Cambridge would be published some years later, partly through the Royal Society, and more permanently in the two volume *Course of Lectures on Natural Philosophy*, which has been reprinted in 2002. The full title of the first paper read to the Royal Society was "Outline of Experiments and Enquiries Respecting Sound and Light", which showed the beginnings of the work on Optics for which scientifically he has become most famous. With regard to sound, the paper shows acquaintance with earlier work by Bernoulli, Euler and Lagrange.

Towards the end of 1797 Thomas Young's great-uncle Dr Brocklesby died. In the will Thomas Young was left his library, his house in London, a selection of paintings mainly chosen by Sir Joshua Reynolds and £10,000. In spring 1799, having fulfilled the residence requirements for his degree in Cambridge, he set up his medical practice in London. Then in 1803 he graduated M.B., and in 1809 M.D., both from Cambridge, while the F.R.C.P. qualification followed also in 1809.

Thomas Young as Professor at the Royal Institution

The founding of the Royal Institution in London must be credited mainly to Benjamin Thompson (1753-1814), a native of New England. He had fought on the British side in the American War of Independence. After coming to this country in 1776 he received a government post before going to live in Bavaria in 1784 as Minister of War and Police. It was while supervising the boring of cannon in Munich that he formed the view that the work done by the horses on the boring machine was transformed into heat. From this observation he concluded that work and heat were different forms of what came to be called energy.

As well as being both a soldier and a scientist, Benjamin Thompson was also a man with a social conscience. The story is told that the multitude of beggars in Bavaria had long been a public nuisance. In one day he caused 2600 of these outcasts to be arrested by military patrols and transferred to an industrial establishment which he had prepared for them. In this institution they were housed and fed, and they not only supported themselves by their labour but earned a surplus. The principle on which their treatment was based was stated by him as follows:

> "To make vicious and abandoned people happy," he says, "it has generally been supposed necessary first to make them virtuous. But why not reverse this order? Why not make them first happy, and then virtuous?"

For his work in this field he was appointed in 1791 a Count of the Holy Roman Empire, with the title of Count Rumford. The name of Rumford was taken from the town in New Hampshire, now called Concord, to which his wife belonged.

After some further years spent in Bavaria and England he proposed in 1799, along with Sir Joseph Banks, who was the President of the Royal Society, the establishment of the Royal Institution, "for diffusing the knowledge and facilitating the general introduction of Useful Mechanical Inventions and Improvements, and for teaching, by courses of Philosophical Lectures and Experiments, the Application of Science to the Common Purposes of Life". The scheme was approved by the King, and a single appointment to a Chair of Chemistry and Natural Philosophy was made.

In 1801 the Chair of Chemistry and Natural Philosophy was separated, with Sir Humphry Davy being appointed to the Chemistry Chair and Thomas Young to that of Natural Philosophy. Thomas Young began his lectures in January 1802 and delivered thirty-one by the end of May. In the following year he increased this number to sixty. Anyone who knew the author would know that the lectures were bound to be scholarly rather than popular.

When they were eventually published in 1807 they occupied two volumes, which have been considered to be so seminal that they were reprinted in 2002. But there is no doubt that, as has been said correctly, the lectures presumed on the knowledge and not on the ignorance of his hearers. Of all the lectures the most significant are those on Optics, which, as has been stated, form part of Electricity. Thomas Young's contribution to Optics will now be considered.

Thomas Young and the Wave Theory of Light

Light rays from any source, from one point of view, may be thought of as carrying minute particles or corpuscles to an observation point such as the eye. Alternatively they may be thought of as waves spreading out from the source in a similar way to sound waves from a sound source. Both ways of looking at such light rays existed even among the Greeks. In Thomas Young's time the corpuscular theory held sway, because Isaac Newton (1642-1727) had, slightly hesitantly, taken his stand in favour of it. In support of the corpuscular theory was the fact that light waves were not able, it was wrongly believed, to bend round obstacles, whereas sound waves could do so. In other words light rays were believed to propagate only rectilinearly.

As far as reflection of light from a plane surface is concerned, both the corpuscular and wave theories can explain this with no advantage to either theory. But when refraction from a plane surface is concerned, the wave theory is simpler, provided one accepts that the velocity of light in a denser medium, such as water, is less than in air. The corpuscular theory, on the other hand, has to assume that the velocity of light in water is greater than in air. No measurements of this velocity were available at this time to provide a clear answer on behalf of either viewpoint.

In the absence of such experimental evidence for light, it was to be expected that arguments based on analogies with other kinds of waves would be considered. It will be recalled that while Young was still at Gottingen he had carried out research using sound waves. It was natural, therefore, that initially Young would consider these. In fact later studies by Fresnel from 1817-1819 were to show that water waves would give a better analogy. And in 1801 Young himself did propose the principle of interference in water waves in the following clear way:

> "Suppose a number of equal waves of water to move upon the surface of a stagnant lake, with a certain constant velocity, and to enter a narrow channel leading out of the lake. Suppose then another similar cause to have excited another equal series of waves, which arrive at the same channel, with the same velocity, and at the same time as the first. Neither series of waves will destroy the other, but their effects will be combined; if they enter the channel in such a manner that the elevations of one series coincide with those of the other, they must together produce a series of greater joint elevations; but if the elevations of one series are so situated as to correspond to the depressions of the other, they must exactly fill up those depressions, and the surface of the water must remain smooth: at least I can discover no alternative, either from theory or from experiment.
>
> Now I maintain that similar effects take place wherever two portions of light are thus mixed; and this I call the general law of the interference of light."

From the penultimate paragraph it is clear that Young performed experiments with his water waves to satisfy himself of this Principle of Interference. It is not certain if he carried out the corresponding experiments with light, but others have, and confirmed the results he foresaw.

In 1817 Young further believed that there might be what he called a transverse vibration in the light wave as well as a longitudinal vibration, which was known to exist in sound waves. Today the word *vibration* would be replaced by *electric field*, which exists only in the transverse direction, and not at all in the longitudinal direction. But the consequences of this were worked out independently by Augustin Jean Fresnel (1788-1827), who will be considered later in this chapter, after we deal with the work of Davy, Oersted, Ampere, and Biot-Savart.

Humphry Davy (1778-1829)

Humphry Davy's grandfather Edmund was a builder, and his father, Robert, a skilled woodcarver in the neighbourhood of Penzance in Cornwall. Grandmother Davy had a vast collection of Cornish stories and legends, including those of the supernatural. The house she lived in was said to be haunted. Humphry as a small boy remembered these stories. He was physically precocious, walking at nine months. He could recite little prayers at age two, and made up verses at age five. Associated with these gifts he had an excellent memory.

At the first school he attended his progress was so rapid that he transferred to Penzance Grammar School at age six. There the master in charge did not like Humphry's attitude, so that it was outside school, among his fellow pupils, that he developed his talent for storytelling. The sources of these stories were partly those he had heard from his grandmother, but he was also familiar from his own reading with *Aesop's Fables* and the *Pilgrim's Progress*. He was aware of local industry, including tanning, so that he could not but be familiar with an introduction to chemistry. In his own extrovert life this took the form of making fireworks and a mixture he called thunder powder, which on explosion would clearly impress his schoolboy audiences. But on a larger scale there were several of James Watt's steam engines with separate condensers operating in Cornwall, to provide power for the tin and copper mining operations. It is believed that unsuccessful speculations in these fields were responsible for a family debt of about £1300 at the time Humphry's father Robert died in 1794. Humphry himself had left Truro Grammar School a year earlier, after spending only one year there, and had gone to live with Dr Tonkin, a family friend. And in February 1795 Humphry became apprenticed to Dr Borlase, a surgeon-apothecary in Penzance. Dr Tonkin's expectation was that Humphry would become a general practitioner, but Humphry himself had higher objectives of graduating in Edinburgh and reaching the top of the medical profession. He was also to maintain two other interests as an apprentice, those of the field sports of fishing and shooting, and composing poetry. This poetical interest will now be considered further.

Humphry Davy and Poetry

The first published poem by Davy was included in Coleridge and Wordsworth's *Annual Anthology* for 1799, though it is believed to have been written in the first year of his apprenticeship in 1795. To provide an indication of its poetic worth, only its first verse is included here:

> "Like the tumultuous billows of the sea
> Succeed the generations of mankind;
> Some in oblivious silence pass away,
> And leave no vestige of their lives behind."

It was good enough to be published, and therefore Davy deserves credit for it.

At this stage it is necessary to say something about the relationship between Davy and Coleridge. They were born in the adjacent counties of Devon and Cornwall, with Coleridge the elder by six years. In 1798 Coleridge had written the first part of "Christabel", and in 1799 in a letter from Coleridge to the poet Southey he refers to it as "the poem with which Davy is so delighted". This statement is illuminating because "Christabel" is a poem about the supernatural. It is referred to in *The Concise Oxford Dictionary of English Literature* as "one of the most beautiful in English poetry", and elsewhere along with "Kubla Khan" and "The Ancient Mariner" as together the three best poems of Coleridge. It deals with the beautiful Christabel praying at night in the wood for her betrothed lover. Behind an oak she finds Geraldine, beautiful also, but a malignant supernatural being. Geraldine says she has been abducted by men on horseback, and Christabel takes her home to her father. He is enchanted by Geraldine and orders a procession to announce her safe return. The poem in unfinished, but it is believed that an outline finish was prepared for it, as given in Wikipedia.

It is likely that Davy's delight in hearing Coleridge read out to him the first part of Christabel in 1799 stems from Davy's own background of the supernatural in Celtic Cornwall. In Celtic Perthshire there is the village of

Aberfoyle, beside which is the Hill of the Fairies. Of this hill a story is told by Robert Kirk, the Minister of Aberfoyle, in his book *The Secret Commonwealth of Elves, Fauns and Fairies*. Robert Kirk was a seventh son, and his successor the Rev Dr Grahame, informs his readers that as Mr Kirk was walking on a Fairy Hill he sank down in a swoon, which was taken for death. Later the form of the Rev Robert Kirk appeared with the message that he was not dead but a captive in Fairyland. In *The Secret Commonwealth* it is stated that "it was a land with aristocratic Rulers and Laws, but no discernible Religion, Love or Devotion towards God, the blessed Maker of all. The inhabitants disappear whenever they hear his Name invoked, or the name of Jesus, (at which all do bow willingly, or by constraint, that dwell above or beneath within the Earth (Phil. 2 v.10), nor can they act ought at that Time, after hearing of that secret Name."

Davy continued to write his own poetry and three verses written by him in 1825 four years before he died are given below:

"And when the light of light is flying,
And darkness round us seems to close,
Naught do we truly know of dying,
Save sinking in a deep repose.

And as in sweetest, soundest slumber,
The mind enjoys its happiest dreams,
And in the stillest night we number
Thousands of worlds in starlight beams;

So may we hope the undying spirit,
In quitting its decaying form,
Breaks forth new glory to inherit,
As lightning from the gloomy storm.

Humphry Davy as Chemist

Humphry Davy is known, correctly, primarily as a chemist and not as a poet. It has been stated that in 1795 Humphry was apprenticed to a surgeon-apothecary. This work necessarily involved practical chemistry in Penzance.

But to study the subject in a more useful way he required a good textbook. Fortunately Lavoisier's *Elementary Treatise on Chemistry* had been published in French in 1789, and this was well studied by Davy. In later years Davy was to refer to him as "the great Lavoisier", remembering that Lavoisier had been guillotined in 1794.

Also in Penzance Davy had the good fortune to meet Davies Gilbert (previously Davies Giddy), previously a Sheriff of Cornwall and later a Member of Parliament. He was much older than Davy, and possessed a library which he made available to the surgeon-apothecary apprentice. Either in one of these books or in one of Dr Borlase's, Davy came across a reference to Nitrous Oxide, on which he started to do experiments, because he disagreed with the author's views that this gas was the cause of contagion. The experiments were shown to Dr Beddoes, a friend of Davies Gilbert, who was setting up a Pneumatic Institution near Bristol, funded in part by Thomas Wedgwood. As a result Humphry, in the fourth year of his apprenticeship, was offered the post of establishing a laboratory to investigate the value of gases in the treatment of disease. Davy accordingly set out for Clifton in October 1798 to work under Dr Beddoes.

Humphry Davy in the Pneumatic Institution at Clifton

Davy began in 1799 to investigate the physiological effect of gases, and their possible benefit in the treatment of disease. Supporters of the project included James Watt and Thomas Wedgwood both of whom had family health problems. Davy's descriptions of the effects of nitrous oxide include the following:

> "A thrilling, extending from the chest to the extremities, was almost immediately produced. I felt a sense of tangible extension highly pleasurable in every limb; my visible impressions were dazzling, and apparently magnified. I heard distinctly every sound in the room, and was perfectly aware of my situation."

After another experience:

> "He often felt great pleasure," he said, "when breathing the gas alone, in darkness and silence, occupied only by ideal existence; and when he breathed the gas after excitement from moral or physical causes, the delight he felt was often intense and sublime."

There were, of course, dangers in these experiments. Davy tried inhaling the gas after drinking a bottle of wine in less than eight minutes. The result was that in less than an hour he sank into insensibility and remained thus for two hours. On the other hand both Bolton and Watt, who were possibly more phlegmatic, were hardly moved at all after inhaling the gas.

Davy in one of his notebooks noted the possibility of the gas removing physical pain from operations. However, he did not patent this idea, and it was only in 1844 that the dentist Howard Wells in America extracted one of his own teeth using nitrous oxide as an anaesthetic.

It appears that sometime in 1799 the potential usefulness of scientific discoveries attracted the attention in London of Thomas Barnard and Count Rumford. This led to the formation of the Royal Institution of Great Britain in 1800, after it received its Charter from George III. Davy was first appointed Assistant Lecturer there in Chemistry in 1801 The Pneumatic Institution under Dr Beddoes did not last long after Davy's departure, partly because Dr Beddoes changed his interests to other fields.

Humphry Davy at the Royal Institution

Davy was an inspiring lecturer and it was not long before he was promoted, first to a full Lectureship and then to the post of Professor of Chemistry at the Royal Institution. This appointment lasted for twelve years until 1814, when Davy was aged thirty-six. The quality of his lectures has been described as follows:

> "The sensation created by his first course of lectures at the Institution, and the enthusiastic admiration which they obtained, is at this period scarcely to be imagined. Men of the first rank

and talent—the literary and the scientific, the practical and the theoretical, blue stockings, and women of fashion, the old and the young, all crowded—eagerly crowded the lecture room. His youth, his simplicity, his natural eloquence, his chemical knowledge, his happy illustrations and well conducted experiments, excited universal attention and unbounded applause. Compliments, invitations and presents were showered upon him in abundance from all quarters, his society was courted by all, and all appeared proud of his acquaintance."

Humphry Davey's Entry into Electricity and After

When Volta communicated in 1800 to the President of the Royal Society his discovery of what has become known as the Voltaic Pile, the information was quickly passed in England to the publisher William Nicholson and his surgeon friend Anthony Carlisle. Within a matter of weeks they had reproduced the same results. But they also observed that oxygen and hydrogen were evolved at the two electrodes. This result intrigued Davy who then began his own experiments in 1800 also. He first found that if the solution of brine, used originally by Volta, was replaced by pure water, that no current is produced. To produce the current the conducting fluid must be able to oxidise the zinc. This discovery enabled Davy to design new types of Voltaic Pile.

Following his appointment at the Royal Institution in 1801, Davy was subsequently required to give lectures on the practical subject of tanning, to make analysis of rocks and minerals, and to give lectures on the chemistry of agriculture. These were all subjects in which he had to educate himself, after he had gathered knowledge of the existing ways in which these things were presently done. Consequently he was unable to continue his electrical researches until 1806. But during one holiday break in 1805 he had the pleasure of meeting Sir Walter Scott, with whom he shared personal and literary interests.

In 1807 Davy discovered potassium by passing an electric current through fused caustic potash. His joy at the success of this experiment has been recorded as follows:

"When he saw the minute globules of potassium burst through the crust of potash, and take fire as they entered the atmosphere, he could not contain his joy—he actually danced about the room in ecstatic delight; and some little time was required for him to compose himself sufficiently to continue the experiment."

Likewise by passing the current through fused caustic soda Davy produced the element sodium. It is recorded that in his Bakerian Lecture of that year Davy "exhibited metals such as mankind had never seen before; metals which swam on water in a molten condition, decomposing it and releasing hydrogen which burned with a beautiful glow, lilac in the case of potassium, and golden in the case of sodium."

Davy was now about thirty years of age, and passing through a restless time. Some friends suggested he should enter the church. They pointed to the example of the Bishop of Llandaff, author of *Chemical Essay*. Davy himself did register at Cambridge so as to enter medicine, but his registration was in absentia, as was permitted at the time. But then an invitation came in 1810 from the Royal Dublin Society to lecture on "Electrochemistry", for a fee of four hundred guineas. The lectures were so well received that he was given five hundred guineas, and in the following year he was awarded the degree of LL.D by Trinity College, Dublin.

In 1812 Davy married Jane Apreece, a wealthy widow whose father owned sugar plantations in the West Indies. The complete dedication of Davy to science probably meant that he would have required an extremely understanding wife. But she did accompany him on travels abroad in the years ahead. Some lines that Davy had written in 1805 were probably relevant:

"No uniformity in life is found:
In every scene varieties abound;
And inconsistency still marks the plan
Of that immortal noble being,—Man,
As changeful as the April's morning skies,
His feeling and his sentiments arise."

Shortly before his marriage Davy was knighted by the Prince Royal.

Still in the year 1812, Davy almost lost the sight of one eye while working with nitrogen trichloride, a chemical more powerful than gunpowder. As a result he was off work for at least two months. One result of this was a decision to employ an assistant to help him, and the person appointed was Michael Faraday, about whom more will be said in the following Chapter. This year of 1812 also saw a coal mining tragedy near Sunderland, in which ninety-two miners were killed. Davy responded to this tragedy by inventing the Davy Safety lamp, for which again he did not apply for a patent. Nevertheless the coal mining proprietors subscribed a dinner service of Gold Plate, which was valued at £2500. In his will Davy requested that this should be melted down to provide a medal for an outstanding piece of chemical research.

Now it is necessary to return to the year 1800 to introduce Hans Christian Oersted.

Hans Christian Oersted (1777-1851)

Hans Christian Oersted was the son of a pharmacist who in 1776 bought a deserted pharmacy on the island of Langeland in Denmark. From his first marriage eight children were born, of whom the two eldest were Hans Christian and Anders Sandoe. Both were to achieve eminence in different fields. Hans Christian was a scientist and Anders Sandoe a politician. During their early years they spent much of their time in a German wigmaker's home while their parents were occupied elsewhere. In that home the wigmaker taught the boys religion from a German bible, and his wife taught them from a Danish commentary. For learning arithmetic, help was provided by neighbours plus the benefit of self-study. Greek, Latin and drawing tuition were available from local sources also.

Since only eighteen months separated the boys in age they were able to study at the same pace. But they also practised teaching each other after a tutor gave individual lessons to one of them. In this way they both learned French and English. When Hans Christian was twelve both boys became apprentices in their father's pharmacy, but after a time Anders Sandoe transferred his interest to the study of Law. Five years on in 1794 they both moved to Copenhagen University, Hans Christian to study pharmacy and his brother to continue with the study of law. When he graduated in 1797 Hans

Christian's marks were the highest recorded up to that time for that university degree.

After graduation Hans Christian became associated with one of the most respected pharmacies in Copenhagen. But in his spare time he interested himself in Electricity, carrying out experiments on the Voltaic Pile in 1801. He also lectured at Copenhagen University, having an appointment as Assistant Professor, but without pay. He then obtained a travelling scholarship for three years which enabled him to work in Germany and France with the best scientists in these countries. It was at the Ecole Polytechnique in Paris that he saw that students were provided with equipment for carrying out their own experiments.

Oersted's first major publication was in the field of acoustics, for which he won a medal from the Royal Danish Society in 1808. He became Secretary of this Society in 1815, and this gave him direct access to the King who was a patron of the Society. But the work for which Oersted is most famous is his discovery of the action of an electric current on a magnetic needle, which he observed in 1820. It is reported that as far back as 1801 he believed that as the luminous and heating effect of an electric current goes out in all directions, he thought it possible that the magnetic effect would also irradiate. To verify this in 1820 he arranged an electric current to pass through a thin platinum wire which was placed over a compass covered with glass. The magnetic needle was slightly disturbed, so little that he did not repeat the experiment for three months. On this second occasion, presumably with a larger battery, the disturbance was more definite but still small, because he employed very thin wires, thinking that the effect would occur better when the wire was luminous. Then he found that conductors of greater diameter gave much more effect. Within a few days he proceeded to discover the circular nature of the magnetic field around the wire carrying the electric current. And that this magnetic field passed through all solids with the exception of iron.

The first discovery following that of Oersted was obtained by Ampere (1775-1836). He was a completely honest man who if he did not like a meal said so, and when bored showed this boredom oozing out of every pore. He was endowed with a phenomenal memory so that in old age he could recite sections of Diderot's Encyclopaedia which he had read as a boy. And when, on his deathbed, a friend started to read from Thomas à Kempis's *Imitatio Christi*, he said, "Thank you but I know the book by heart." This was the

Ampere who within five years of having heard of Oersted's discovery then developed a complete mathematical theory of the electromagnetism of steady electric currents.

Thus two theories of electromagnetism now co-existed, the non-mathematical one of Oersted, and Ampere's. When Oersted visited Ampere several times in April 1822, Ampere went out of his way to convert Oersted to his point of view. He even invited scientific friends along to dine with Oersted, among them Fourier and Fresnel, and discussions lasted on one occasion for three hours after the meal ended. But Oersted was able to demonstrate that his non-mathematical theory fully accounted for all the known phenomena, leaving both theories valid.

Andre-Marie Ampere (1775-1836)

The storming of the Bastille as the first act of the French Revolution in 1789 ushered in a period of political terror in that country. It included the execution by guillotine of the French King Louis XVI in 1793, and likewise those of the political activists Danton and Robespierre in 1794. These events took place in Paris, but the instability and terror existed also in the industrial city of Lyons, the home of the silk industry, where it is said 30,000 people became unemployed at that time. There in Lyons, Jean-Jacques Ampere, the father of Andre-Marie Ampere, who was earlier associated with the silk industry, and later was a justice of the peace and president of the police tribunal, was also guillotined in 1793. This followed the death in the previous year of Andre-Marie's elder sister, who had been his constant companion. In later years he would continue to write poetry in her memory. The combination of these two deaths had a shattering effect on Andre-Marie Ampere. For over a whole year at age eighteen he did nothing. When he did recover somewhat at age nineteen it was to start writing poetry. His father had been his only teacher and he had early access to his father's library. As a boy he delighted in memorising whole passages, and in particular memorised the whole of *The Imitation of Christ*. For the Ampere family devotion was considered more important than dogma. It has been recorded by Andre-Marie's son that his father's first communion was one of the three most important events of his youth. The other two were a biography of Descartes which led him into science, and the fall of the Bastille which determined his political allegiance.

But in 1796 Andre-Marie met the woman, Julie Carron, he was to marry three years later. She was the daughter of a successful silk merchant in Lyons, and for the four years of their married life he was happy. But illness took her away. Before she died he entered the French educational system, giving tutoring in mathematics, chemistry and languages. Then in 1801 he became Professor of Physics and Chemistry at Bourg, and in 1804 Professor of Mathematics at the Lycee in Lyons.

In the same year of his appointment at Lyons a Society of Christians was formed there, of which Ampere became President. It had seven members originally and each of them was given a theological proposition to investigate. The title of the one Ampere was given was "To present an exposition of the historic proofs of Christianity", i.e. basically to show that Christianity was divinely inspired. This he did by providing three proofs. First from the Old Testament, then from enemies of Christianity, and finally from Christian writers who had recorded the miracles of the disciples and Jesus himself. But these were intellectual exercises and Ampere himself struggled with his religious doubts to achieve both an intellectual and a spiritual equilibrium. An aphorism attributed to Ampere is, "Doubt is the greatest torment that man has on earth."

In 1804 also Ampere was appointed as repetateur in the famous Ecole Polytechnique. A repetateur was a tutor for students who received lectures from the Professor. This meant leaving Lyons for Paris, and it was in Paris he stayed for the rest of his life. He himself was promoted to Professor in 1815 and taught continuously at the Polytechnique until 1828.

During his time in Paris Napoleon was crowned Emperor. This was followed by the initial restoration of Louis XVIII in 1814, then by Napoleon's return for one hundred days, then by the second Restoration, and by the Revolution of 1830. These changes of political power did not greatly affect Ampere, though Napoleon did not win his respect. It is of interest that in the Restoration period there was a Catholic revival in France.

The reputation of Ampere as a scientist was achieved primarily through his work at the Polytechnique in the field of electromagnetism, which will now be considered.

Ampere and Electromagnetism

After having seen a replica of Oersted's experiment on the mechanical force exerted on a magnetic needle by a conductor carrying an electric current, Ampere wrote, "When M. Oersted discovered the action which a current exercises on a magnet, one might certainly have suspected the existence of a mutual action between two circuits carrying currents; but this was not a necessary consequence; for a bar of soft iron also acts on a magnetised needle, although there is no mutual action between two bars of soft iron."

Therefore Ampere carried out experiments with steady currents, and showed

(1) wires carrying electric currents exert mechanical forces on each other,
(2) mechanical forces are exerted on wires by magnets.
(3) two wires close together carrying equal currents in opposite directions have a negligible resultant magnetic effect.
(4) a wire twisted into any shape has the same magnetic effect as if it were straightened out.
(5) the mechanical force exerted on a wire carrying a current is always normal to the wire.
(6) the mechanical force exerted by one short current element on another varies as the inverse square of the distance between them, just as is the case for the force between two electric charges.
(7) the magnitude of the mechanical force exerted on a short current element placed in a magnetic field is proportional to the size of the current, to the length of the element and to the strength of the magnetic field. Ampere also derived the direction of this mechanical force as being the vector product of current direction and magnetic direction.

These results were all obtained within about five years after Oersted's seminal experiment. Ampere was able to express the results in a quantitative way because of his mathematical skills and he deserves the title of being called the Father of Electromagnetism. This primacy in science, however, was not accompanied by happiness in his personal life. His second marriage was unhappy, the son from his first marriage was a disappointment to him, and his last years were spent in poverty. But, looking back, he should always be remembered with admiration.

Jean-Baptiste Biot (1774-1862) and Felix Savart (1791-1841)

These two contributors following the discovery of Oersted that a magnetic field surrounded a wire carrying an electric current will be considered together. This is because Biot, the senior of the two, recruited Savart to help him make measurements. Biot was a member of the Laplace School at the Ecole Polytechnique where he had graduated in 1797. At that time Laplace was preparing *Mecanique Celeste* for publication, and because Biot wanted to study this great work in advance, he offered to proofread the text. At times he would ask Laplace to explain steps which had been skipped over with the phrase, "It is easy to see." It is good to have been told by Biot that sometimes Laplace would forget how he had obtained a particular result and had difficulty in reconstructing it. But it is also good to know of the relaxed atmosphere which Laplace cultivated with his young research workers. After a morning of intellectual work they would sometimes have lunch together with Madam de Laplace, a lunch of frugality with milk, coffee and fruit. During this get-together they would discuss scientific problems for hours on end, with Laplace suggesting future research work for them, but also discussing their future prospects.

When Biot was a student at the Ecole Polytechnique he took part in an insurrection by the Royalists and was taken prisoner. Only the intervention by the Professor Monge of the Ecole Polytechnique saved him from possible death. Later in 1804 he took part in a balloon ascent to a height of 13,000 feet to study the magnetic, electrical and chemical composition of the air at various altitudes. This showed that the earth's magnetic field does not vary significantly up to this height of 13,000 feet. As a champion of the corpuscular theory of light he devised ingenious explanations for polarisation.

Felix Savart's family came from an engineering background, but at the age of seventeen in 1808, he decided to go in for medicine. After two year's study he qualified as a surgeon in Napoleon's army. Following Napoleon's defeats in Spain and Russia he was discharged from the army and resumed medical studies. Then after graduation he spent much time on a translation of *De*

Medicina by Celsus, who lived around the beginning of the Christian era. His *De Medicina* was long one of the chief manuals on medicine. Savart then in 1819 went to Paris to obtain a publisher for his translated *De Medicina*, and also to meet Biot who could discuss with him his other interest in stringed musical instruments. One of these constituted a design for a trapezoidal violin which Savart claimed to have superior acoustic performance to the traditional violin. A committee assessment of it reported as follows:

"When the instrument was played before a committee that included Biot, the Italian composer Cherubini and other members of the Academy of Sciences and the Academie des Beaux-Arts, its tone was judged as extremely clear and even, but somewhat subdued."

When Savart arrived in Paris he found Biot engaged in research on electricity as well as acoustics. In the following year they collaborated in producing the Biot-Savart Law which says:

"The magnetising force, at any observation point, due to an electric current I flowing in a circuit C, is the same as if each element of C contributed a vector proportional to the size of the current I and length of the element, and inversely proportional to the square of the distance from the element to the point of observation."

Experimentally they knew that the first thing they had to discover was how the law of the force due to the conducting wire decreased in strength at various distances from its axis. To make their measurements as precise as possible they neutralised the magnetic effect of the earth by using an appropriately located bar magnet. Their magnetised needle was then subjected to the magnetic effect of a long vertical conducting wire at various measured distances.

Augustin Jean Fresnel (1788-1827)

Augustin Jean Fresnel was born in 1788 in Broglie, about fifty miles south east of Caen. Most of his childhood was spent in the village of Mathieu nearby. Up to the age of twelve he was taught at home by his parents. His father was an architect who undertook major projects such as improvements in the Chateau of the Duke de Broglie, and the construction of Cherbourg Harbour.

This home instruction from his parents was within the religious environment of the Jansenist movement. Jansenism was a sect within the Roman Catholic Church which had been founded by Cornelius Jansen (1585-1638), Bishop of Ypres, on the basis of his book *Augustinus*, a bulky treatise on the theology of Augustine (354-430).

Theological Views of Jansen

Jansen believed that the Church suffered from three evils at that time. The first evil was that the official scholastic theology was anything but evangelical. Secondly, simple souls found spiritual pasture only in little mincing "devotions". Thirdly, more robust members built a religion quite as close to the views of Epictetus as to Christianity.

Epictetus (b. c. A.D.60) recognised that there is only one thing that is fully our own, i.e. our will. God has given us a will which cannot be compelled or thwarted by anything external. We are not responsible for all the ideas that come to us, but for the way we use them. There are natural instincts within us of self-preservation and self-interest. But man is so constituted that he cannot secure his own interests unless he contributes also to the common welfare. In this way man will grow into the mind of God.

Jansen thus advocated religious experience in place of reason, love of God in place of churchgoing, and the realisation of the helplessness of man in place of moralistic self-sufficiency. For this to be achieved Jansen saw the necessity of conversion. But this depended on God's good pleasure, and this implied predestination.

These views naturally were opposed to the views of the Jesuits, whom Jansen accused of giving absolution much too readily, and of an excessive reliance on sacramental grace. But he also opposed the views of Protestants. Partly this was because the distinction between Justification and Sanctification had not been fully worked out by the Protestants at that time. As a result it was obvious that some people who claimed to have been justified were not leading sanctified lives. The other main reason he opposed the Protestant church was that he believed that a personal relationship between a man and his Maker could only be obtained through the Roman Catholic Church. Nevertheless the overall Jansenist viewpoint was not accepted by that Church. It persecuted the Jansenists and eventually suppressed them in 1731.

Two of the most famous scientists associated with Jansenism in France are Blaise Pascal (1623-1662) and A. J. Fresnel whom we are now considering.

Fresnel's Education

It has been said that for the first twelve years of his life Fresnel was educated at home by his parents within the Jansenist background of that home. A saying well known within the Jesuit community is that given a boy for the first seven years of his life he will remain a Catholic for the remainder of his life. It is clear that a similar result applies to Fresnel.

At age twelve Augustin Jean Fresnel was transferred to the Ecole Centrale in Caen. Here he was introduced to science and mathematics, and after four years he moved to the Ecole Polytechnique in Paris. At this Institution his maternal uncle already taught Design. Among the staff he was introduced to was Ampere who was newly appointed, and among the students was Arago, both of whom were later to become colleagues and friends.

After studying at the famous Ecole Polytechnique for two years Fresnel progressed to the Ecole des Ponts et Chaussees, where he trained as a Civil Engineer for a further three years. Thereafter he was engaged in different projects, including the building of a major road connecting Spain and Italy, beginning in 1812. But it appears that in addition to practical engineering he had scientific and philosophic interests as well. From his family upbringing he wanted to question the truths of some of the beliefs presented to him as matters of faith. Even in science he had been taught the corpuscular theory of light. One of the consequences of this theory was that light travelled only in straight lines. But obstacles provide evidence of light bending round them into the shadow. Fresnel, like others, did not understand this.

It was in 1815 that political events took a turn which enabled Fresnel to give up his civil engineering responsibilities. Napoleon had escaped from the island of Elba, and Fresnel had enlisted in what turned out to be a small force to oppose his return to power in France. But Napoleon prevailed and Fresnel was dismissed from his post. Consequently Fresnel returned to his home village of Mathieu.

To begin with Fresnel needed experimental results which would allow him to measure the positions of the light fringes produced by obstacles. To achieve this he needed a micrometer. He also needed a heliostat to provide light rays from the sun which would arrive in a fixed direction. Since he had neither of these he was forced to construct them himself. It is on record that a key advance was made when he fixed a piece of black paper to one edge of a diffracting obstacle and found that the bright bands within the shadow vanished. This showed that the bright bands had been coming from both edges of the obstacle.

Within a few months of arriving in Matthieu, Fresnel had submitted two extensive memoirs to the Academy of Sciences in Paris. There they were assessed by a panel which included Arago. As a result Fresnel was authorised to spend some months in Paris at the beginning of 1816 in order to repeat his experiments with better equipment. This he did, obtaining substantially similar results.

But Fresnel was also a mathematician and worked out formulas which gave the positions of the bright and dark lines which he had obtained in his experimental results. This was a major achievement, but it regrettably duplicated results which had been obtained earlier in Cambridge between 1797 and 1799 by Thomas Young. It so happened that Fresnel knew no English. There was further disappointment for Fresnel in France itself, because Laplace and Biot still remained convinced supporters of the corpuscular theory. They arranged for an Academie Prize on Diffraction to be announced, with submission of the *Memoirs* by 1818. In this connection there is a very human story told of a communication from Ampere to Fresnel relayed by a mutual friend:

> "Yesterday I saw Ampere, who asked me for news of you and strongly enlisted me to write to you to put yourself in the ranks and to send your memoir to the contest, with the new observations that you have made and that you may yet make. 'He will assuredly win the prize,' he said to me, 'for himself and for the cause he must compete.' I made some objections, based on the partiality of the commissioners if they were chosen from the sect of the Biotistes—Ampere replied that there was

nothing to fear; that when the commissioners were nominated, General Arago would not fail to make known the impropriety of nominating party men, and that what will happen is what always happens when the Republic is warned that citizen Laplace wishes to dominate."

Arago rose to the occasion; he and Gay-Lussac joined Laplace, Biot and Poisson on the commission that judged the entries for the Diffraction Prize. Fresnel was declared the winner.

There was one interesting comment from the commissioner Poisson. From Fresnel's analysis Poisson was able to make a prediction of an effect which had never been observed experimentally. This applied to the case of a circular obstacle illuminated along its axis. Poisson showed mathematically that everywhere on the axis of this circular obstacle there would always be a bright spot of light, the opposite of what one might expect. This was then confirmed experimentally.

Further credit is due to Fresnel for discovering that light waves behave more like water waves than sound waves. This is because the amplitude of the electrical vibration in the light wave is perpendicular to its direction of travel. In contrast the pressure of a sound wave is along its direction of travel. It is to the credit of Fresnel that he records that the possibility of this transverse vibration came to him from Ampere, although the discovery of its being completely transverse belongs to Fresnel.

Physical illness attended Fresnel for much of his life, and he died of tuberculosis at the age of thirty-nine. During the last five years of his life he stayed in Ampere's house. Ampere was a devout Roman Catholic and Fresnel a devout Jansenist; yet they collaborated as friends and colleagues. Fresnel died in the arms of his mother, who said:

"I thank God for giving to my son the grace of using the talents he received for their usefulness and the general good. For everyone to whom much is given, from him much will be required, and to whom much has been committed, of him they will ask the more." (Luke 12 v48.)

CHAPTER SIX

M. FARADAY : THE SANDEMANIAN INFLUENCE

(c.1826 – c.1880)

THE CONTRIBUTORS TO THE ADVANCEMENTS in electricity described in Chapter Five came from England, Denmark and France. In this Chapter pride of place for the continuing advancements shifts, initially chronologically, to Germany, England and America. It is of interest to note the importance of high quality journals in Germany with regard to the rise in the national rankings. *The Annalen der Physik und Chemie* was founded in 1790 by F. A. C. Gren, Professor of Physics and Chemistry at Halle University. He wrote, "the mind requires theories of natural phenomena that bring unity and cohesion to our conception of them." After his death in 1798 the editorship passed to L. W. Gilbert, a colleague at Halle, of Huguenot descent, who spoke French, English, Dutch and Italian as well as German. In his editorship the title of the Journal became simply *Annalen der Physik*, and it achieved international recognition before he died in 1824. Thereafter Johann Christian Poggendorff became Editor, and it is to him we owe a correction to the work of Georg Simon Ohm (1789-1854), who then gave us the correct form of what we now know as Ohm's Law, published in 1826.

Chronologically the next momentous discovery in electricity was made in England by Michael Faraday (1791-1867). He shared Oersted's non-mathematical approach to electricity. It was a view that electricity, magnetism, light, heat and even chemical reactions are caused in some way by similar forces, without specifying what these forces were. This fitted in with Faraday's theological beliefs, in which the laws of Nature shared similarities, resembling, but not being identical with, the different works of a single sculptor or painter. Thus electricity, magnetism, light, heat, chemical reactions and all the different forms of life could be interpreted as the varied output of a single Creator. This, as far as the different forms of life are concerned, would form an alternative interpretation to that of an evolutionary development. But Faraday also believed in creation out of nothing, and within non-living creation electricity represented for him the most magnificent of God's powers. He stood in awe of it. It emphasised the vast gap between God and man, when a comparison was made between the vast electrical powers in a thunderstorm, for example, and the small powers then available from a Voltaic pile.

In parallel with some of Faraday's researches was the work carried out by Joseph Henry (1707-1878) in New York. His grandparents were of Scottish origin, and Henry lived with his grandmother after his father died while Henry was still young. His mother has been described as a woman of great refinement and a Presbyterian of the old-fashioned Scottish stamp, of a deeply devotional character. She is reputed to have exacted from her children the strictest performance of religious duties. In Joseph Henry's case this sense of duty was transmitted to his choice of career, by rejecting wealth to become a scientist. He also refused to patent his inventions, believing that their benefits should be freely available for all. Two of his many inventions included, in 1831, a powerful, iron-cored electromagnet which could lift 750 lbs., equal to thirty-five times its own weight. Another was a reciprocating rather than a rotating electric motor. Further, as an example of a scientific discovery, in 1832 he and Faraday independently discovered self and mutual induction. On the basis of prior publication Faraday received the honour for mutual induction and Henry for self-induction.

Returning now to Germany we consider the contribution made to electricity by Wilhelm Weber (1804-1891). He was one of twelve children born to a Professor of Theology at Wittenberg. Both he and two of his brothers, Ernst Heinrich Weber and Eduard Frederich Weber, became noted scientists, each of them working in the medical field. In Wilhelm's case it is possible that the influence of his father may have been responsible for the son's joining with six other Professors at the University of Gottenberg in 1837 in protesting against the revocation of Hanover's liberal constitution by the newly appointed ruler, the uncle of Queen Victoria. As a result all seven were dismissed, including the two Grimm brothers.

William Weber was a man of broad talents. At age twenty-one he co-authored with his brother Ernst a book of 575 pages on wave motion, which gave the first detailed application of the hydrodynamic principles involved in the study of the circulation of the blood. Eight years later he co-authored with his other brother Eduard a book on the physical mechanism of walking. In between these two publications he was invited by Carl Friedrich Gauss to be a Professor at the University of Gottingen. It was from Gottingen that Gauss organised the setting up of worldwide observatories for recording the earth's magnetic field variations, and Weber was his main experimental collaborator until his dismissal in 1838. Further from 1832 onwards until 1846 Weber worked on Electrodynamics. In this period, which included his time at Leipzig as Professor of Physics there, he revolutionised Electrodynamics. This he did by considering the mechanical forces on electric charges within conducting wires, a completely basic approach. In the course of this theory he came close to obtaining a theoretical value for the velocity of light. Finally in 1871 he advanced the view that atoms contain positive charges surrounded by negative charges which migrate from one atom to another.

The final component in this group of five seminal developments starting from about 1825 relates to Franz Neumann (1798-1895). His father was a farmer who moved into estate management. In this new capacity he served under a Countess who had been divorced. But her parents would not allow the marriage of their daughter to one of a lower social class. Consequently Franz Neumann was brought up by his father's parents, and did not meet his mother until he was ten years old. What may be a related consequence of this is the remark of Hermann von Helmholtz (1821-1894) who after meeting Franz Neumann in 1848 said of him, "He was somewhat difficult to approach, hypochondriachal, shy, but a thinker of the first rank."

It is recorded, though sometimes disputed, that it is to Franz Neumann that we owe the introduction in current electricity of a new type of potential. Unlike the potential associated with electric charges or magnetic poles, both of which are simple numeric quantities, this new potential has both magnitude and direction associated with it. It is therefore called a vector potential. Earlier potentials were defined as the work done in bringing unit charge from infinity to the point at which the potential was to be found. The work done was positive because a force of repulsion had to be overcome all the way as the unit charge was brought from infinity to the point where it was required to know the potential.

In the case of vector potential since like currents attract each other, it is necessary to consider oppositely directed currents to obtain a positive force of repulsion as one brings up a short current element from infinity to the point where the vector potential is to be found. The advantage of introducing this vector potential is that when a current changes, the size of an induced electromotive force due to this change of current is given by the rate of change of the vector potential. It may be to Franz Neumann that we owe this discovery. Moreover, the direction of the electromotive force in space, called more simply the induced electric field, is the same as the direction of the vector potential.

Each of these contributions to the greater understanding of electricity will now be considered in turn.

Georg Simon Ohm (1789-1854)

Georg Simon Ohm was the son of a locksmith, and his mother the daughter of a tailor. He was born in Erlangen in Bavaria, a town which owed its prosperity to the influence of French Protestants after the revocation of the edict of Nantes by Louis XIV in 1685. The original Edict had been signed in that city by Henry IV in 1598, giving Protestants the right to the free exercise of their religion and opening to them all offices of State. After its revocation about 400,000 Protestants left France for Britain, Holland and other Protestant countries. Erlangen Protestant University was founded in 1743, and even today the Protestant tradition is so strong that there are six Protestant churches, and only one Roman Catholic Church in Erlangen.

Father Ohm gave his sons a sound introductory training in Mathematics, and also in the Philosophy of Kant. After the age of eleven Georg Simon Ohm attended the Erlangen Gymnasium, and at age sixteen entered the University of Erlangen. Here his attendance lasted for only three semesters because, it is believed, of his abstemiousness with regard to study, among other available activities. Consequently his parental support was withdrawn and Georg emigrated to rural Switzerland. Here he supported himself by teaching and tutoring for almost five years, before being advised by his former Mathematics tutor in Erlangen to study Euler, Laplace and Lacroix on his own.

Because of his innate mathematical ability he was able to do this and returned to the University of Erlangen at the age of twenty. He spent four years there, obtaining his doctorate in two years and then spending the remaining years as a Research Fellow. For whatever reason, possibly associated with his family problem or because of his own personality, Ohm did not continue in University life. Instead he became a mathematics and physics teacher at a secondary school in Bamberg, about thirty mile south of Coburg. He stayed there for four years, and it was there he published his first book, at age twenty eight, entitled *Geometry as a means of Education*. Ohm believed that "the student should learn mathematics as if it were the product of his own mind, and not be something that was imposed from outside". This emphasis on the centrality of the individual contrasted with the philosophy of the contemporary philosopher Georg Wilhelm Friedrich Hegel (1770-1831). It was, however, a view shared by his younger brother Martin Ohm who became a Professor of Mathematics at the military academy in Berlin. Here he too had to face opposition from the education authorities, and acquired the reputation of being a revolutionary. But this view was to be extended still further in Denmark through the writings of Soren Kierkegaard (1813-1855) who strongly opposed the philosophy of Hegel. It will be helpful here to say something more about Hegel because of later problems encountered by Georg Simon Ohm with Hegel's administrative supporters in Berlin.

Georg Wilhelm Friedrich Hegel (1770-1831)

Georg Hegel was born in Stuttgart, the son of a Secretary in the revenue office. It may be assumed that Georg was able to learn his own later habits of orderly work from observation of his father's activities. His mother was the daughter of a lawyer who worked at the High Court of Justice in Stuttgart, but she died when Georg was only thirteen. This bereavement may have contributed to his precocious gravity. By race, which was one of the five instruments of education which Hegel was later to promulgate, Hegel himself was a Swabian, but despite that a Protestant in religion, which was a second instrument of education in Hegel's beliefs. The other three instruments were the family, the school and the people or nation.

At age five he went to a Latin school for two years, and then transferred to the Stuttgart Gymnasium (a classical school), where he spent the next ten years. His progress there was excellent, within the confines of the teaching system used. This appeared to concentrate on note taking, from poets and dramatists, including the German dramatist Lessing, all of which Hegel accomplished in a very orderly fashion. However this may have been responsible for his early belief that his later life should concentrate on being a populariser of philosophy rather than being a cutting-edge philosopher.

After leaving the Gymnasium Hegel became a theology student at the University of Tubingen from 1788-1793. Because he was destined for the church he held a bursary at the Protestant Theological Seminary. But he became so absorbed in his own studies that he missed lectures. And at the end of the course he decided to become a teacher, because he believed this would provide more time for his own study interests. The teaching posts which were available to him initially took him to Switzerland where he spent the years up to 1799, the year in which his father died. Having been left some money from his father's estate Hegel delivered lectures at the University of Jena, as a research student, after submitting a dissertation on the orbits of the planets. Later he became an extraordinary Professor; but both were unpaid positions. This occupied the years 1801-1806. In the year 1806 Napoleon defeated the Prussians at the battle of Jena, and the city was devastated. Consequently Hegel was again out of work until he was offered the post of Editor of a newspaper in Bamberg. This work he undertook for one year until he was offered and accepted the double post of Professor of Philosophy and Rector of the Nurnberg Gymnasium.

Hegel believed that all work done for the State was worthy of a man's best efforts. He regarded education as a State function in which the teachers were civil servants. In the Gymnasium the number of teaching hours was twenty-seven per week, and homework was to be added to this. Written reports were made on each pupil by the master who taught him, and these were read out in each class so that the work and conduct of each boy was judged in the presence of his classmates. In one of his annual addresses to pupils leaving the Gymnasium, Hegel said, "Youth looks forward; in doing so, however, never forget the backward look of gratitude, of love and duty towards your parents." He also did his best to ensure that his pupils should not only attend public worship with their parents on Sundays, but arranged that they should keep the religious festivals of their respective churches.

During his time in Nurnberg Hegel was also writing his great book, the *Logic*, and this was published in 1816, the year he left the Gymnasium. His fame had grown to the extent that in that year he was offered three University Chairs, at Erlangen, Heidelberg and Berlin. His philosophy had developed to the extent that he wished to lecture to more mature minds. He accepted Heidelberg for two years and then moved to Berlin in 1818 where he stayed until his death in 1831. Of his own teaching he said, "I am a schoolmaster who has to teach philosophy, and perhaps partly for that reason, am possessed with the idea that philosophy, as truly as geometry, must be a regular structure of ideas which is capable of being taught." He exercised an important influence over Prussian education in Berlin, partly because some of his former students occupied administrative office there. In addition his philosophical system was the one officially acknowledged in the Ministry of Education and Medicine's decree commanding the Royal Science Examinations Commission to suppress any "shallow and superficial" non-Hegelian philosophy.

What that philosophical system said is illustrated in two quotations from Hegel in the Oxford Dictionary of Quotations:

> "Civil society has the right and duty of superintending and influencing education, inasmuch as education bears upon the child's capacity to become a member of society. Society's right here is paramount over the arbitrary and contingent preferences of parents."

and,

> "Only in the State does man have a rational existence.—Man owes his entire existence to the State, and has his being within it alone. Whatever worth and spiritual reality he possesses are his solely by virtue of the state."

Of course it should not be thought that Hegel's views were universally accepted. Even in his own time his early work was criticised by Carl Gauss, who said of his doctoral thesis, "Noah got drunk only one time, to become then, according to the Scriptures, a judicious man, while the insanities of Hegel where he criticises Newton and questions the utility of a search for new planets are still wisdom if one compares them with his later remarks."

Likewise a century later George Santayana (1863-1952), denying the view that might makes right, said, "The worship of power is an old religion, and Hegel is full of it.—Such a master in equivocation could have no difficulty in convincing himself that the good must conquer in the end if whatever conquers in the end is the good."

Georg Simon Ohm (contd)

The publication of Ohm's book *Geometry as a Means of Education* in 1817 could hardly have come at a worse time for the author, when Hegel's fame was approaching its zenith. Recalling that Ohm believed the student should learn mathematics as if it were the free product of his own mind, while Hegel believed the student's mind should be guided by the requirements of the State, meant that officialdom in Berlin would ignore his views. Nevertheless the book must have helped his teaching progression somewhat, despite the administrative opposition to it. As a result Ohm was appointed in 1818 a senior teacher of Mathematics and Physics at a Jesuit Gymnasium in Cologne. This school had a well-equipped laboratory, and after Oersted's discovery of electromagnetism in 1820, Ohm was able to do experimental work in both electricity and magnetism. In parallel with these his theoretical studies in mathematics now included the works of both Fourier and Fresnel, in addition to the earlier French Mathematicians.

Within five years of Oersted's discovery, Ohm, at age 36, submitted a combined experimental and theoretical article for publication, to Poggendorff the editor of the *Annalen*. It is to Poggendorff we owe the suggestion that Ohm should repeat his experiments using a thermoelectric source in place of a voltaic cell. The terminal voltage from the voltaic cell decreased with time, so invalidating the previous results. Ohm accepted the suggestion despite having to use a lower voltage, and was then able to submit a corrected paper to Poggendorff. This was published in 1826.

In 1826 also Ohm applied for leave of absence from his post at Cologne so that he could concentrate on research. This application was referred by the Prussian Minister of Culture to Professor Erman, the Professor of Physics in Berlin who was regarded by Laplace as the foremost physicist in Germany. This was a fortunate referral for Ohm since there had been warfare in departments of Philosophy in Germany between those who were content to collect specimens and those who required new experimental equipment. It has been reported that Erman "looked in anger on attempts to explain Prussia's military defeat by Napoleon through intellectual renewal by means of idealistic philosophy". The result of the referral was that Ohm was granted leave of absence for one year on half salary, and he moved to Berlin where his younger brother was a Professor in the Military Academy and had a house there.

At the end of the year's leave of absence Georg Simon Ohm published his large Monograph entitled *Galvanic Circuits*. With regard to current electricity he proved two things:

(1) The quantity of electricity transferred from one body to another was proportional to the difference between their potentials and inversely proportional to their distance apart.
(2) Heterogeneous bodies in contact maintain a constant difference in potential across their common surface.

Ohm was very careful in his assessment of what was happening in even a simple galvanic circuit. He did not look at a circuit from a systems point of view in which it is sufficient to say that the current is equal to the electromotive force divided by the total resistance. He also said the current at a point was due to a small difference of potential between adjacent particles of electricity on each side of that point.

The Galvanic circuit did not sell well. This was partly due to there being too few people who could follow his mathematics. It is reported that Ohm did his best to increase the sales by asking friends to purchase copies outside Berlin so as to impress the publishers. A copy was sent to the Prussian Ministry of Education. Along with it Ohm resigned his position in Cologne and expressed the hope of obtaining a more appropriate position. Unfortunately this did not materialise. And in addition a reviewer of his monograph, perhaps a supporter of Hegel who believed that reasoning made experimental work unnecessary, wrote the following cutting commentary on the publication: "He who looks on the world with the eye of reverence must turn aside from this book as a result of an incurable delusion, whose sole effort is to detract from the dignity of nature."

For six years until 1833 Ohm served as a private tutor in Berlin. Then he was appointed as a Professor at the technical college in Nuremberg, but not at a university. However in 1841 he was awarded the Copley Medal of the Royal Society in London, and finally at age 60 in 1844 he was appointed Professor of Physics in Munich, dying five years later.

Michael Faraday (1791-1867)

Ohm was handicapped by the philosophical beliefs of Hegel. Faraday, on the other hand, was strengthened by the theological beliefs of the Sandemanian Church in which he had been brought up. This denomination was started by John Glas (1695-1773) who was educated at St Andrews and Edinburgh Universities before becoming the Church of Scotland Minister at Tealing, a few miles north of Dundee. In Scotland the denomination is known as the Glassite Church. Robert Sandeman was the son-in-law of John Glas and was responsible for the later growth of the Church in both England and the United States.

It was within the predominantly supportive life of this Church that Faraday spent his working life. It affected his response to the outside world. It provided him with the assurance that there was a distinctive plan behind events. If he fell sick his fellow Church members rallied to his support. When visiting British Association meetings outside London he would, where possible, join in their Sandemanian or Glassite services, either as a member of the congregation or as a speaker. He spent every Sunday and other occasions during the week at the Church meetings. At the Sunday meetings the communion bread was broken together, and at other meetings they washed each other's' feet. The Love Feast in the middle of the day on Sundays was shared with fellow members together in a separate room, while non-members and children were provided with separate refreshment in the Church pews. Sandemanianism for all its church members was a way of life. They did not, however, seek to extend their influence by missionary work to outsiders. They simply tried to do good to all men.

Because of their absence of missionary work it is perhaps not too surprising that the Sandemanian Church no longer exists today. The last existing meeting house in Britain, which was situated in Edinburgh, is understood to have closed in 1989. But it therefore lasted for one hundred and ninety years after its doctrinal basis, which had been largely formulated by Sandeman, and had been demolished in 1789 by Andrew Fuller, a Baptist Minister in Kettering. Thus Faraday, who was a boy of eight at that time, spent virtually the whole of his life in a church without a logical intellectual foundation. But, of course, it had a very real spiritual value within its own membership. As a result of this it could ignore the illogicalities pinpointed by

Fuller, such as the theoretical possibility of an impenitent believer worshipping within its fellowship.

With regard to the organised structure of the Sandemanian Church there were three stages in its hierarchy. First there were ordinary members for whom the only requirement was attendance at worship. Secondly there were members who had made public confession of their sin and also profession of their faith in Jesus Christ. The third group consisted of those who had been elected to the eldership in the Church. This involved being able to assist in the worship and sometimes to conduct services, since there were no full-time pastors associated with the sect.

Michael Faraday was aged thirty when he moved from being an adherent to being a member of the Church. This was one month after his marriage to Sarah Barnard, the daughter of an elder in the London congregation. She was already a member, and when she upbraided him for not having told her in advance of his intention, he simply replied, "That is between me and my God." It was a further twenty years before he was asked to become an elder. This request caused spiritual turmoil, bringing to mind the consciousness of his own spiritual unworthiness. But gradually this was replaced by a recollection of the many promises of God, which bring comfort to His people, and he accepted.

And so in 1840 Michael Faraday became an elder in the Sandemanian Church. But by this time he had become famous because of both his chemical and electrical researches. So famous that a story is told of his being invited to lunch by Queen Victoria and his acceptance of that invitation. Since the lunch was on a Sunday it meant having to miss the service in the church, and being unwilling to show repentance for his action, he was removed both from the eldership and also from church membership. It was almost seventeen years later before he was restored to the eldership, by which time he was sixty-seven years of age. Probably due to ill health he resigned from this office after a further four years.

James Kendall's biography *Michael Faraday: Man of Simplicity* draws attention to a sequence of alternate events in Faraday's life, in which periods of high achievement appear to be followed by events of unhappiness. Thus his first discovery in electromagnetism resulted in a charge of having failed to give proper acknowledgement to another scientist, Wollaston, who had attempted unsuccessfully to carry out a slightly similar experiment. There were also disagreements of a similar kind with his predecessor Humphry

Davy. Such disagreements are not uncommon in the scientific world, especially when experimental work is involved. It is less common in theoretical work where a unique type of mind is required to produce something that is characteristic of that type of mind.

Michael Faraday's Early Life

Michael Faraday's grandfather Robert was a tenant farmer. The size of his holding was under fifty acres, and he operated a small mill driven by a stream which passed through his land. There were ten children, of whom seven were boys. This meant that the boys had to acquire trade skills, since the produce of the land could support only one family. The third son, James, who was to become Michael's father, became apprenticed to a blacksmith. Eventually he set up his own business at Kirkby Stephen in Westmoreland. This business initially did well, but the effect on trade of the French Revolution in 1789 was such that he decided to move to London on the south side of the River Thames. Here their son Michael was born in 1791. James's health was not good and for many years he was unable to work for the length of a full day. Eventually he died in 1810.

An elementary school provided education for Michael. Of this education he is recorded as saying later, "It consisted of little more than the rudiments of reading, writing and arithmetic at a common day school. My hours out of school were passed at home and in the streets."

Michael Faraday's Life as an Apprentice

At age thirteen, in 1804, Michael began work as a message boy for a bookseller and bookbinder. After one year he became an apprentice in this firm which belonged to a very kindly owner. There were other apprentices too, and an early indication of Michael's competitiveness can be seen in his pride that he could strike one thousand blows with his mallet pounding the book pages without resting. Another indication of his determination to use his time profitably in life was his practice of early morning walks in London to observe what had been built architecturally, and to see what grew around him botanically.

But the major advances in his years of apprenticeship developed from his reading. These need to be covered under the headings of General, Chemical and Electrical reading. Under General reading he benefited greatly from the writings of Dr Isaac Watts. For most people, Isaac Watts is thought of only as a Hymn Writer, the author of, for example, "O God our help in ages past." But most of his hymns were written in the early years of his adult life. After that he became a tutor to a noble family and later a pastor. At age fifty-one he produced a book on Logic which became a standard text in universities both in England and the U.S.A. In the following year he produced a scholarly work on Astronomy. One of Isaac Watt's books on *Improvement of the Mind* came to Michael Faraday's attention during his apprenticeship. In it a recommendation to keep "A Commonplace Book" recording ideas and interesting facts was taken up by Michael under the title "The Philosophical Miscellany". Also attendance at lectures was recommended and he began in 1810 to attend lectures at the City Philosophical Society, on different branches of science.

Under Chemical reading he had access to the four-volume *Thomson's Chemistry*, but much more important was a new book *Conversations on Chemistry* published in 1805 and written by Mrs Jane Marcel. The book was written on the basis of Humphry Davy's lectures at the Royal Institution which started in 1802. Jane Marcel herself was not a chemist, but was determined to explain chemistry in a straightforward way. She did this by inventing three characters, a teacher and two girl pupils. The clarity of her writing was such that sixteen editions of the book were published, and more than one hundred and sixty thousand copies of the book were sold in the U.S.A. alone. In order to verify the statements in the book Michael Faraday conducted experiments himself as a check on their accuracy. As a further indication of this book's value the Royal Institute of Chemistry has made available since 2005 copies of its tenth edition of the book, originally published in 1805. Mrs Marcel was a prolific author and wrote also *Conversations on Political Economy, Conversations on Evidences of Christianity, Conversations on Vegetable Physiology*, and *Conversations on the History of England*.

Under Electrical reading Michael Faraday had access to the *Encyclopaedia Britannica*. But at that time the article on Electricity dealt with Electrostatics only. To obtain information on current electricity he attended about a dozen lectures by a Mr Tatum between 1810 and 1811. These cost a shilling per lecture, and this was paid for by Michael's elder brother Robert, also a blacksmith and the breadwinner of the family at the time of their father's final illness. At these lectures Michael met Benjamin Abbot, a Quaker who was to become a lifelong friend, who was working as a clerk in the city.

The year 1812 was to be an important year in Michael Faraday's life. His apprenticeship ended in October of that year, but before that date he had the opportunity of attending four lectures by Sir Humphry Davy at the Royal Institution, between February and April. This had been made possible through the generosity of a customer, Mr Dance, in his employer's bookshop, who had been shown copies of notes and drawings made by Michael of lectures by Mr Tatum which he had attended. The four lectures he was now to attend at the Royal Institution were to transform his life. He now wished to give up his bookbinding skills and spend his life in what he assumed would be the high moral ground of science. To this end he applied to the President of the Royal Society for any position in Science, however menial. It came to his notice that his letter did not merit a reply.

But in October of 1812 Sir Humphry Davy's eye was injured by an explosion when he was working with nitrogen trichloride. As a result he was unable to read or write. Possibly through the recommendation of the customer Mr Dance, Michael Faraday was asked to be Sir Humphry's amanuensis. And for a few days he carried out this work. In late December 1812 Michael then wrote directly to Sir Humphry asking for a position in Science, and including the leather-bound volume of the notes he had made of the four lectures by Sir Humphry he had attended. But there was no vacant position—until an altercation between two Royal Institution employees resulted in one being dismissed, and Michael was offered the position of Assistant to Sir Humphry there. The salary was to be one guinea per week plus accommodation at the top of the Institution building.

By April 1813 Michael experienced four different explosions when assisting Sir Humphry with further work on nitrogen trichloride. In one explosion he lost part of a nail and the glass mask he was wearing for the protection of his face was cut. But by September a new opportunity arose. Sir Humphry wanted to meet his continental fellow scientists despite the war with France which was then ongoing. But Napoleon was a supporter of Science. The French Institute in 1807 had awarded Sir Humphry the prize created by Napoleon for outstanding work in galvanic electricity. So Napoleon was pleased to allow Sir Humphry's party to travel in France.

Consequently Michael Faraday had the opportunity of travelling with the party, initially as Assistant to Sir Humphry, and later additionally, though regrettably, as valet. The party embarked in October 1813 from Plymouth, when Michael was aged twenty-three, having never been more than twelve miles out of London before.

On 23rd November their party was called on in Paris by Andre Marie Ampere, who was to become famous in electricity seven years later, after the discovery by Oersted of the magnetic effect of an electric current. Among the visitors along with Ampere were two chemists who had purified a solution obtained from burned sea algae by adding sulphuric acid to it. To their surprise a purple vapour arose which condensed into dark crystals with a metallic lustre. A sample of this was provided to Sir Humphry who after further tests pronounced it to be a new element to which he gave the name Iodine, because it had the form of a violet, for which the Greek root is *ion*

The next most scientifically interesting place they visited was Florence, where the Grand Duke owned a large lens which could be used for igniting diamonds. Previous experiments had produced the result that the only product of combustion had been carbon dioxide, indicating a composition of carbon alone. But graphite, looking quite different, also appeared to be composed of carbon alone. After further tests in both Florence and Rome Davy accepted both results as being simply indications of different crystalline forms. From Rome to Naples and Mount Vesuvius they turned north and in Milan met Volta; then to Geneva, Zurich, Munich, Florence again, Rome again, and Naples again. By this time news arrived that Napoleon had escaped from Elba, and Sir Humphry decided to return home in April 1815. When he did so, he arranged for Michael's promotion with an increase of salary to thirty shillings per week.

The years from 1815 to 1820 were spent on chemical researches in which Michael Faraday began to take an increasingly leading role. In 1820 itself Sir Humphry became President of the Royal Society, but this event was far overshadowed by Oersted's discovery of the magnetic field which accompanies every electric current. The details of that discovery were communicated to Faraday much later, by Professor Hansteen of Christiania in Norway, who had witnessed Oersted's actual discovery. It has been described as a discovery which Oersted "tumbled over by accident," or more kindly as partly serendipitous.

The year 1821 saw the marriage of Michael Faraday to Sarah Barnard, the daughter of a London silversmith, who was also a Sandemanian. She was aged twenty-one and Michael was thirty. There is no record of a honeymoon after the marriage, and the couple had no children.

The same year saw Michael writing a historical review of electromagnetism. To do this he not only read the appropriate research papers but also checked on authors' experimental results, and carried out such additional experiments as he considered necessary. This review was published in *Annals of Philosophy*. Such a review was good not only for those who might read it, but also for the author himself. In this case it actually led to the invention of the first electric motor. Looking at this with the benefit of hindsight we now know that there will be a mechanical force exerted on any conductor which is carrying an electric current when it is placed in a magnetic field. This is what Faraday arranged experimentally, almost certainly on the basis of trial and error. But it produced joy at the time, even though it would be many years before it would become a commercial product.

It is, however, a fact of experience that periods of joy are often followed by periods of dismay. It is inevitable that they cannot ever be followed by continuous growing joy. Thus they must be followed by a levelling out, or a fall in joy which may be sufficiently deep to be called dismay. This is what happened to Michael Faraday.

It is necessary here to draw attention to a failed experiment by Dr Wollaston and Sir Humphry on April 21 1821, in which Dr Wollaston believed that a long current-carrying conductor would be rotated when a bar magnet approached it at right angles to the wire. Michael Faraday was not involved in the experiment but passed through the room when it was being carried out. The experimental result was that no rotation of the wire took place. When Michael Faraday's own paper was published he was accused of stealing Dr Wollaston's idea. There is no evidence that Dr Wollaston felt strongly about the incident but it is clear that Sir Humphry Davy did not express in public, as he should have done, his complete trust in Michael's integrity. Other examples between 1821 and 1824 of friction between Humphry and Michael were associated with the liquefaction of gases and also the election of Michael as a Fellow of the Royal Society.

Nevertheless Michael's reputation continued to grow. In 1823 he was a founder member and the first Secretary of the famous Athenaeum Club, with Sir Walter Scott another founder member. Then in 1824 he was appointed to a Chair of Chemistry at the Royal Institution at a salary of £200 per annum, an appointment for life. The following year he isolated benzene, which has been described as the greatest chemical discovery of his life, being the basis of all aromatic compounds in dyes, perfumes and medicinal products. But during most of the 1820's it appears that he abstained from work in electromagnetism so as not to clash with Sir Humphry, after the priority problem of the electric motor discovery. When in 1829 Sir Humphry died, Michael discontinued his work, which at that time was on glass, and returned to electricity. In 1822 he had written in his notebook, "Convert magnetism into electricity." In the intervening years he would have known of the work of William Sturgeon, who in 1825 found that an iron horseshoe, varnished for insulation, and wound with a coil through which an electric current was passed, could lift a weight of nine pounds. And then in 1830 Joseph Henry using nine separate coils in parallel achieved a lifting weight of six hundred and fifty pounds.

Michael Faraday's Contribution to Electromagnetism in 1831

It has been pointed out by Professor Cramp in his survey of Faraday's work given at the Faraday Centenary Celebrations in 1931, that the order of the work recorded in his Notebooks is different from that given in the book *Experimental Researches*. He also points out that the Notebooks show how the mind of Faraday went step by step, and allow us to see into his mind at the time.

It appears that the first experiment Faraday carried out, used a toroidal iron ring of about six inches outside diameter and of height seven eighths of an inch. Since his object was to produce electricity from magnetism it was natural to start with an iron core which would, from his knowledge of electromagnets, produce a stronger magnetic field than an air-cored coil. On this iron ring he wound separate copper wire coils each twenty-four inches long, which could be connected separately or be joined together. One coil was connected to a battery and a galvanometer was connected across one of the other coils. When Faraday connected or disconnected the battery a flickering deflection was observed on the galvanometer. He was probably hoping for a continuous deflection but this was not obtained.

Then instead of magnetising the ring with one coil and with the second coil closed he brought a permanent magnet up to the ring. He was unable to see any electrical effect. Can the effect be due to using a ring? The ring was replaced with an iron bar with two coils and resulted in the same transient effects as with the ring. He then repeated the experiment of bringing a permanent magnet to the second coil which was closed, and again there was no result. He then said, "All the effects are due to the electric current"— which was a false conclusion due to insensitive instruments.

During the week 24-31st September 1831 he had one positive result when he used a short coil with a cylindrical iron core four inches long connected to his galvanometer, and moved the coil between two bar magnets. The success was immediate and he declared, "Magnetism has been converted into electricity." When the iron core was removed the experiment failed, again due to insensitive instrumentation. But when he put two coils of copper round a block of wood, each coil having thirty-four turns each of seventy-three inches in length, with the turns alternating with each other, he found there was a strong inducing effect.

On 17th October 1831 he took a hollow helix of eight coils of copper wire, each containing twenty-seven and a half feet of copper, and connected the eight coils in parallel to the galvanometer. Into the solenoid he plunged a cylindrical bar magnet which produced a deflection of the galvanometer needle. When the magnet was stationary the needle returned to its first position, and on being withdrawn the deflection was in the opposite direction.

He then tried a disc moving between magnet poles with contacts at the centre and outside edge of the disc, these contacts being connected to a galvanometer. This failed, but he thought it should work and availed himself of a large magnet belonging to the Royal Society. This worked and on 4th November 1831 he enunciated the following rule: "If a terminated wire moves so as to cut a magnetic line, a power is called into action which tends to urge an electric current through it."

Faraday's results were read on 24th November 1831. Two sets of phenomena are described in it. The first is that a varying current in one coil produces a transient current in a second coil. This phenomenon underlies every transformer, every coupled circuit and every induction coil. The second phenomenon is that relative movement between a magnet and a coil or disc produces a current. This underlies every dynamo, alternator and electric motor.

Approximately concurrent with the work of Michael Faraday was the work of Joseph Henry in America, and to that we shall now turn.

Joseph Henry (1799-1878)

Joseph Henry was born in Albany, the capital of the State of New York, in 1799. The name Albany was taken from the title of the Duke of Albany, who was later James II of England and VII of Scotland. The title was generally given to the younger son of the King of Scotland, with its derivation coming from the name Alba, which is the Gaelic name for Scotland. The town of Albany was originally acquired from the Dutch in 1664 and their name for it was Fort Orange.

The Henry family were immigrants who had arrived in New York City in 1775. At that time their name was spelled Henrie and they came from an Argylshire family distantly related to the Earls of Stirling. It was known that other Scots families had settled in the region of Albany after the Battle of Killiecrankie in 1715, and so it was not surprising that the Henrie family should head to farm in that area on arrival from Scotland.

Joseph Henry was the grandson of the immigrant William Henrie, who lived to age ninety. But William's son, another William, was not strong and died in Joseph's early years, so that Joseph's parental upbringing was mainly from his mother. She has been described as a woman of great character, intelligence and refinement. She was of a deeply religious nature and exacted from her children the behaviour of a Scottish Presbyterian kind. Probably because of her husband's ill-health Joseph was transferred to his grandmother's home at Galway, about thirty-five miles from Albany, when he was aged seven. This was a rural community, and Joseph attended school there for three years. At the age of ten he started work part-time in a local store, but spent the rest of the time in school, keeping this practice up for five years. During this time Joseph developed the practice of being the centre of attention among his peers. He was able to do this through the knowledge he had acquired from books plus the gift of a good imagination. In this respect his practice was much like that of Humphry Davy in his days in Penzance. It may be that a Celtic connection between Argyllshire and Cornwall may be at the root of this common practice in their youth.

At age fifteen Joseph returned to Albany and joined a dramatic company for one year. But his interests were then transformed through the reading of a copy of Dr Gregory's lectures on Experimental Philosophy, Astronomy and Chemistry, published in London in 1808. The copy was made available to him at age sixteen by Robert Boyle, a fellow Scot who was one of Mrs Henry's boarders in Albany. The book had a conversational style similar to that which Faraday found so useful in *Conversations in Chemistry*. He thereafter took two years of evening general study in English and Mathematics, in which he supported himself by daytime teaching in a district school. This was the only post for which Joseph ever applied, since every other post later in life was offered to him without applying for it. He was then recommended at age eighteen to a position of being private tutor to the family of a patron in Albany. Since this involved only morning work it allowed the opportunity of private study for the two years of the appointment. This time was used to good effect through the reading of Lagrange's *Mecanique Analytique*. It gave a quantitative completeness to his study of mechanics, though Joseph did not himself become a mathematician.

Because there were few opportunities for employment in the field of science, Joseph Henry decided to prepare himself for a career in medicine. He therefore attended lectures in anatomy and physiology, and began to assist the Principal of Albany Academy in preparing his chemical lectures and demonstrations. These lectures drew large audiences in a city which had two scientific societies, later amalgamated into one, with about two hundred and fifty members. Joseph Henry was appointed Librarian of this Albany Institute, and gave his first lecture "On the Chemical and Mechanical Effects of Steam" to the Institute at age twenty-six.

It appears that the effect of his concentrated study was such that a rest was called for. Through the wisdom of a local judge he was then offered a position of surveyor for a projected new road going west from Kingston, about seventy miles south of Albany, to Lake Erie. This was a substantial civil engineering project covering a distance of over three hundred miles. But it was accomplished so successfully within twelve months that he was then offered the supervision of constructing a canal in Ohio. In parallel with this came an offer to accept the chair of Mathematics and Natural Philosophy at the Albany Academy. This had a lower salary than the canal project, but it was the one he accepted, offering more satisfaction by following what he regarded as the path of duty in preparing young people for public works.

The teaching duties began in 1826. There were four professors and one hundred and fifty students. Joseph Henry began the practice in natural philosophy of demonstrating the principles liberally by illustrative experiments, rather than using a blackboard alone. In his teaching duties he would have been familiar with the writings of Gilbert, who investigated electric attraction in several substances, and with von Guericke who invented an electrostatic machine. He would also have known of the Leyden jar, and he had a copy of the Abbe Nollet's book which demonstrated the strength of charge which could be stored in these jars.

Further, he would have known of Franklin's lightning rod and of Priestley's book *History and Present State of Electricity*, published in 1767, which was available in Albany's State Library after 1824. Then there was Galvani's animal electricity, and Volta's battery. And in 1826, the year of his appointment in the Albany Academy, he saw in New York one of the new electromagnets devised by Sturgeon, an English electrician. This was able to sustain a weight of nine pounds.

It was not until 1827 that Joseph was able to start on improvements to this electromagnetic system, and this work continued until 1831. The aim he set himself was to maximise the lifting power while using a small battery. By increasing the number of coils on the electromagnet, with a battery consisting of a single pair of 4 inch by 6 inch plates, he could lift a weight of thirty-nine pounds. Proceeding further with a battery having a single plate of zinc and half a square foot of surface, he made a magnet lift a weight of 750 pounds— more than thirty-five times its own weight.

In 1831 also, Joseph Henry and Michael Faraday were independently working on the problem of how to generate electricity from a magnetic field. Faraday was the first to show how a momentary electric current could be produced by a change of magnetism brought about by a change in a nearby electric current. He also showed how this electric current could be maintained through the mechanical motion of a conductor cutting through lines of magnetic force. But it appears that Joseph Henry preceded Michael Faraday in publishing in 1831 the phenomenon of self-induction. A vivid spark is produced when an electric current is suddenly broken. This result was independently made two or three years later by Faraday, who appears not to have noticed Henry's previous publication.

In 1832 Joseph Henry moved to Princeton as Professor of Natural Philosophy. Here he constructed his largest electromagnet, one capable of holding a weight of 3600 pounds. He also introduced a telegraph system in the following way. A conductor of approximately one mile in length was connected to a battery source at one end. At the other end it was connected to a powerful electromagnet. Close to one pole of this electromagnet was placed a permanent magnet on a pivot so that it was free to rotate. When the circuit was closed and the electromagnet was energised the permanent magnet was repulsed and caused one end to swing sharply against an office bell. Thus an audible signal was produced. It is recorded that later in life when he was asked why he did not patent the telegraph, he answered, "I did not then consider it compatible with the dignity of science to confine benefits which might be derived from it to the exclusive use of any individual." This was then followed by the comment, "In this I was perhaps too fastidious."

It is clear from these two inventions alone that Joseph Henry's reputation was high in Princeton. Consequently it is not surprising that in 1836 the trustees of the College granted him a year's leave of absence on full salary. At that time it was understandable that he should decide to visit Europe. This would require letters of introduction, and for this purpose it was necessary to visit Washington. There Joseph met the President Elect, Martin van Buren, who received him cordially and agreed to provide the necessary letters. Accommodation was booked on the maiden voyage of the *Wellington* which departed on 20th February 1837.

On arrival in London he visited the American Ambassador, who received him helpfully and offered assistance in providing introductions. One such introduction had already been made from a friendship established through the Franklin Institute in Philadelphia. As a result Joseph Henry met a former American Ambassador whose present role was associated with James Smithson's will. James Smithson was the illegitimate son of Hugh Smithson, the first Duke of Northumberland and Elizabeth Macie, a descendant of King Henry VII of England. He had been ostracised in England and lived much of his life in Paris. It appears that at no time had he any direct connection with the United States, though it has been surmised that an American acquaintance in Paris had the idea of making Washington a cultural capital in that country. The relevant part of the will of James Smithson simply specifies that "his fortune should go to the United States of America, to found at Washington, under the name of the Smithsonian Institution an establishment for the increase and diffusion of knowledge among men." This legacy took the form, substantially, of over one hundred and five thousand golden sovereigns. When the Smithsonian Institution was eventually opened in 1845, the Chief Executive, with the title of Secretary, was chosen to be Joseph Henry.

But this appointment was several years ahead. Clearly in 1837 one of his chief hopes during his visit to England was to meet Michael Faraday. Scientifically they were rivals but both were religious men. Because of Faraday's other commitments, Henry had to spend some considerable time with Mrs Faraday, and was able to record a wife's observations on her husband. Since such views are often illuminating the following are included. "He was fond of novels, he was not disposed to go into company, but associates principally with a few persons, denies himself to everyone three days in the week, never dines out, except when commanded by the Duke of Sussex at the anniversary of the Royal Institution."

During this same visit when Michael Faraday was not available, Joseph Henry spent some time also with Charles Wheatstone, who in 1834 was appointed Professor at Kings College in London. He took out a patent in 1837 for an electric telegraph, and is usually regarded in Britain as the inventor of this means of communication. But Joseph Henry preceded him by a few years. On the occasion of one such visit to Kings College when Michael Faraday was present along with John Frederick Daniell, the inventor of the Daniell Cell, the following story is told. An attempt was made to generate a spark from a thermopile, i.e. a thermocouple which converts heat into electricity. All three Englishmen tried unsuccessfully to do this. When Joseph Henry then asked permission to try he inserted a long wire wrapped round an iron core, to increase its self-inductance, and this was enough to produce the desired spark. The story then concludes with the reaction of Faraday, who it is said jumped to his feet and exclaimed, "Hurray for the Yankee experiment!" The relationship between the two rivals was not only not hostile but good.

This first visit to England as part of his European tour lasted for eight weeks. It was a matter of regret for him to leave the friends he had made. His next port of call was Paris and here he was at a disadvantage in not speaking French. He then visited Holland briefly where he found the resemblance to his native Albany very pleasing. After that there was a return to London, where Faraday took him on the first twenty-five miles of the railway to link the capital with Birmingham. A pleasing feature of this trip was being allowed to ride on the locomotive.

After this it was on to Edinburgh and a visit to Scottish lighthouses, which formed an introduction to his later work on American lighthouses. His letters home to his wife on this part of the tour showed a wide knowledge of Scottish history and literature. He was familiar with the poems of Ossian, the real or mythical Irish poet of the second or third century, which may have been worked up into two epic poems "Fingal" (1762) and "Temora" (1763) by James Macpherson (1736-1796). Macpherson studied at Aberdeen and Edinburgh Universities, became a schoolteacher and translator of poems from Gaelic or Erse, and later was a member of the House of Commons and is buried in Westminster Abbey. The relevance of this is that Joseph Henry met a daughter of James Macpherson in the wife of Sir David Brewster (1781-1868), who was famous for his research work on the polarisation of light. In practical work he invented the polyzonal lens for use in lighthouses and the kaleidoscope. But as recounted by Thomas Coulson in his biography of Joseph Henry, Sir David Brewster possessed a handwritten letter by Sir Isaac Newton to a widow of his acquaintance. The letter apparently set forth in a series of semi-mathematical propositions how it was the lady's duty to forget the memory of her departed husband in the arms of the living scientist.

On his return to Princeton in 1838 Joseph Henry continued his electromagnetic work. His first achievement was connected with discharges from the Leyden jar. What he found was that the discharge did not consist of a single restoration of the equilibrium. Instead there was a sequence of discharges back and forth, which gradually diminished to zero. He proved this by passing the discharge through a coil in which needles were placed with different degrees of magnetisation. After the discharge they were found to be magnetised in different directions, showing that going and return currents had passed through the coil in which they had been placed.

He then produced in 1843 an electro-chronograph which was designed to measure the speed of a projectile. It did this by measuring the time of transit between two screens in the path of the projectile, when these screens were a known distance apart. Then in 1845 he carried out measurements on the heat radiated by sunspots. An unusually large sunspot in the centre of the sun's disc had been observed. It was natural to ask the question if this would have any effect on terrestrial temperatures, and Herschel believed that the sunspot would increase this temperature. A telescope was used to produce an image of the sun's disc on a screen in a darkened room. The diameter of the sunspot observed on the screen was about two inches. Then a thermopile was moved across the image of the sun's disc. This showed that the sunspot was cooler than the surrounding regions.

In 1845 also, Joseph Henry was appointed Secretary of the Smithsonian Institution, to which he devoted the rest of his professional life. But it is pleasant to record that he was able to advise Alexander Graham Bell in 1875 with regard to his later working invention of the telephone in 1877. He himself died in 1878.

Wilhelm Eduard Weber (1804-1891)

Returning now to Germany we consider the contribution to electricity made by Wilhelm Weber. He was one of twelve children born to Michel Weber, a Professor of Theology at Wittenberg, where Luther nailed his ninety-five theses to the church door, and where both he and Melanchthon are buried. Of the four sons, one became a Minister, two became Professors of Medicine in Leipzig University and Wilhelm became a physicist and Professor, predominantly at Gottingen University.

Wittenberg was a small town, and the Weber family lived in a house belonging to a Professor of Medicine and Natural History, a house in which also lived a lodger, who was to become famous in acoustics. This was Ernst Florent Friedrich Chladni, (1756- 1827). He investigated the laws of sound and carried out experiments on the vibration of plates of different sizes and shapes. In particular, when a horizontal metal or glass plate covered with sand is clamped at one point and set in vibration by means of a violin-type bow, the resulting figures formed by the sand are known as Chladni's Figures. Clearly, Wilhelm Weber had a privileged upbringing in Wittenberg.

Nonetheless problems came to Wittenberg when it was bombarded by the Prussians in the war against Napoleon in 1813. In the following year the Weber family moved to Halle, about fifty miles south west of Wittenberg. Halle was a much larger town than Wittenberg and the two Universities combined in 1817. Michel Weber obtained a Chair in Theology in Halle.

Wilhelm received his early education from his father, but then attended a grammar type school in Halle before entering Halle University at age eighteen. Within four years he obtained his doctorate in 1826 for a dissertation on the theory of reed organ pipes. Two years later, in 1828, at a conference in Berlin, organised by Alexander von Humboldt, he gave a talk on this subject which attracted the attention of both the organiser, and also of Carl Gauss, the Professor at Gottingen University. The existing interest of von Humboldt in geomagnetism transferred to Gauss, who saw in Weber a suitable experimental co-worker for Gottingen, should a position become available there. It would also have been known to Gauss that three years before this conference Wilhelm Weber had co-authored a book with his elder brother on Wave Theory, which included experimental work on water as well as sound waves. The book also included a study, using hydrodynamic principles, of the circulation of the blood. This gave an impressive indication of his total work capability.

In the same year as the Berlin Conference, Wilhelm Weber was appointed to a teaching position at Halle University. But when in 1831 a position of Professor of Physics at Gottingen University became vacant, Weber moved there to begin collaboration with Carl Gauss. Within one year this produced a joint paper in which the strength of a magnetic property could be reduced to measurements of length, time and mass. This meant that it could be reproduced anywhere. They also started a network of magnetic observatories to correlate the resulting measurements, and introduced a two-mile long telegraph system between their physics laboratory and the astronomical observatory to facilitate simultaneous measurements. Further, in teaching, Weber was responsible for introducing practical experiments for the students, and the friendship with Gauss was such that Gauss started to lecture in Physics, as opposed to the purely mathematical lectures he had given since his appointment in 1817.

In addition to his work in the Physics Department at Gottingen, Wilhelm Weber also managed to find time to collaborate with his younger brother in Leipzig University. Between them they produced a book in 1836 which dealt with the physiology and physics of walking.

The year 1837 saw the accession of Queen Victoria to the throne in Britain. But under Salic law it was not possible for a woman to take over the throne in Hanover. Consequently the rule there passed to Victoria's uncle, the Duke of Cumberland. One of his early actions as King was to impose unconstitutional encroachments on the liberty of the Hanoverians. In protest against this seven University Professors in Gottingen recorded their dissent. They became known as the "Gottingen Seven". Among them were the two

Grimm brothers of fairy tale fame, and Wilhelm Weber. The Duke of Cumberland dismissed all seven from their University posts in 1838, and exiled some of them, including one of the Grimm brothers.

It is relevant to say that at the time of his dismissal Weber was aged thirty-four and was single. In fact, he never married. After the dismissal he travelled in Germany, and to England and France, as a promoter of earth magnetic observatories, and making personal contact with scientists in these countries. On his return it appears that he maintained contact with Gauss, and in addition with his friend Poggendorff, the Editor of the *Annalen* in Berlin. When in Berlin he always visited the mathematician P. G. L. Dirichlet (1805-1859) and the family of Dirichlet's wife, the Mendelssohns, whose home was described as the most distinguished meeting place of arts and sciences in Berlin. Dirichlet himself became Carl Gauss's successor in Gottingen in 1855.

But it was not until 1843 that he succeeded in returning to a Chair of Physics. This came about through the retiral through ill-health of Gustav Theodor Fechner (1801-1887), the Professor of Physics at Leipzig. Fechner is remembered now mainly for his experimental work in Psychophysics in which he advanced the view that the sensation experienced in acoustics or vision is proportional to the logarithm of the stimulus. Since Weber had earlier postulated this view in 1834 the law is now called the Weber-Fecher Law. So in 1843 Weber took up Fechner's Chair in Leipzig. Since his salary here was almost twice what he had received in Gottingen, and he had the promise of a new magnetic laboratory in Leipzig, these compensations must have been welcomed by him.

For the remainder of his professional life Weber worked on the subject of Electrodynamics. He remained at Leipzig only until 1849, by which year political events in Europe allowed a return to a Chair in Gottingen, and collaboration with Gauss again. It is to his credit that at his request his 1838 replacement at Gottingen was retained on Weber's return there in 1849.

Weber's work on Electrodynamics was typically fundamental, although it is largely ignored today. The reason for this is a present-day emphasis on a systems approach to problems in electricity. Thus in practical situations it is adequate to talk or measure the electric current in a circuit using Ohm's Law. But this does not consider what is happening in the conductor carrying the electricity. In that conductor it is the electric charges which constitute the current. These charges are operated on by an electric force causing them to move. Weber considered different components in that electric force. For example, in electrostatics the electric force is due to other electric charges alone in the conductor. But when a constant electric current flows, an additional force is applied to every charge through the total magnetic field which exists at its location. Then, if the electric current is not constant, but changes with time, there is a further additional force which depends on the acceleration of the charges. Thus there are three component forces acting on every electric charge in the conductor. These forces have different magnitudes and different directions, and except in electrostatics, change with time, so that the desire to ignore them all quantitatively and simply use Ohm's Law is understandable.

An additional contribution from the work of Weber was his suggestion in 1871 that atoms contain positive charges surrounded by rotating negative particles. Then he envisaged these negative particles migrating from one atom to another when a voltage was applied to the conductor. Overall his extensive contributions to science were huge, so it is not surprising that he was awarded the Copley Medal of the Royal Society in 1859. In 1935 the name "Weber" was chosen for the unit of magnetic flux, and hence the more commonly used unit of flux density becomes the "Weber per square metre".

Franz Ernst Neumann (1798-1895)

The city of Konigsberg, originally part of East Prussia, and since the Potsdam agreement, capital of Kaliningrad in Russia, is famous for being the birthplace of the Philosopher Immanuel Kant. It is also the burial place of the mathematician F. W. Bessel and the physicist F. E. Neumann.   Franz Neumann's father was a farmer, Ernst Neumann, and his mother was a divorced Countess. Her parents would not permit her to marry Ernst Neumann, so that Franz was brought up by his paternal grandparents. It is recorded that he did not meet his mother until he was ten years old. It may be that this is relevant to a comment made by the famous physicist Hermann von Helmholtz when he first met Franz Neumann, then aged fifty, in 1848. He said, "He was somewhat difficult to approach, hypochrondriacal, shy, but a thinker of the first rank." For comparison his comment on Wilhelm Weber, made three years later, was, "After Neumann he is the first mathematical physicist in Germany."

Franz Neumann attended the Berlin Gymnasium as a boy where he demonstrated particular ability in mathematics. But he left the Gymnasium in 1814 at age sixteen, to join the Prussian army. In this early decision he displayed a firm emotional characteristic which remained with him throughout his life. As examples of this, he entered Berlin University at age nineteen to study theology in accordance with his father's wishes, but left after six months because this discipline was not appropriate for him. At age twenty-five when his father died he gave up a promising career in mineralogy to run the farm now belonging to his mother. Throughout his life he was a highly influential teacher, with many of his students becoming famous scientists in their own right. Then at the age of forty-nine he built a physics laboratory for the benefit of his department with his own resources, when the government were unable to fund it. But this firm emotional aspect was accompanied by an other-worldliness in the sense that he believed priority of discovery extended equally to lectures and publications. Consequently he published only a fraction of his work, as major portions of it were given in his lectures. And it appears that although his lecture notes were prepared for publication by his son in 1895 they did not appear in print. In keeping with Neumann's rather otherworldly attitude is his saying "The greatest reward lies in making the discovery; recognition can add little or nothing to that."

When Neumann ran his mother's farm for a year after his father died, this did not stop him writing his first paper, on mineralogy, at age twenty-five. He then continued his researches at the University of Berlin, taking his doctorate two years later. In 1829 he was appointed to the Chair of Mineralogy and Physics at the University of Konigsberg in East Prussia. There he was influenced by F. W. Bessel (1784-1846), the famous astronomer, and by K. G. J. Jacobi (1804-1851), famous for his work on Elliptic Functions, and his interest changed towards mathematical physics. After four years in Konigsberg, Neumann and Jacobi started a mathematical-physics seminar to introduce students to research. This practice must rank as one of the most outstanding developments in the history of research. It is another example of Neumann's concern for the development of his students.

With regard to Neumann's contribution to electricity, this came later in his life. His first paper in 1845, when he was aged forty-seven, provided the magnitude of the electromotive force in a coil through which the magnetic flux was changing with time. It will be recalled that Faraday's discovery was not expressed by him in mathematical terms. This shortcoming was filled in by Neumann who said that the electromotive force was simply equal to the rate of change of the magnetic flux through the coil.

A later development by Neumann provided another mathematical formula which enables the self and mutual inductance between coils to be found. This is of use in calculating the electromotive force induced in one coil by the changing current in another coil coupled to it.

In 1887 Neumann was awarded the Copley Medal of the Royal Society.

CHAPTER SEVEN

KELVIN: THE AGE OF THE EARTH

c.1850 – c.1900 (Section 1)

THE PERIOD 1850-1900 saw major advances in the understanding in part, of electricity, as are to be outlined in Chapters Seven and Eight. In Chapter Seven the work of the German Hermann von Helmholtz will first be described, followed by that of the Irish-Scot William Thomson, later Lord Kelvin. Then the architectonic theory of the Scot James Clerk Maxwell, later to be the first Director of the Cavendish Laboratory in Cambridge, and the man who was responsible for providing the theory which displaced the continental theory of William Weber, will be outlined. Finally the contribution of the German Heinrich Hertz, who has been described as the star pupil of Helmholtz, will be described. He was the man who produced, for the first time, radio waves in air, thus making radio communication possible.

Hermann Von Helmholtz (1821-1894)

Few scientists or philosophers have earned the respect which came to Helmholtz during his own lifetime. On the occasion of his seventieth birthday in 1891 he gave a speech from which the following extract is taken:

> "I have been overloaded with honours, with marks of respect and goodwill in a way which could never have been expected. My own sovereign, his Majesty the German Emperor, has raised me to the highest rank in the Civil Service; the Kings of Sweden and of Italy, my former Sovereign the Grand Duke of Baden,

and the President of the French Republic have conferred Grand Crosses on me; many academies, not only of science, but also of the fine arts, faculties and learned societies spread over the whole world, from Tomsk to Melbourne, have sent me diplomas, and richly illuminated addresses, expressing in elevated language their recognition of my scientific endeavours, and their thanks for these endeavours, in terms which I cannot read without a feeling of shame. My native town, Potsdam, has conferred its freedom on me. To all this must be added countless individuals, scientific and personal friends, pupils and others personally unknown to me who have sent their congratulations in telegrams and in letters."

Hermann Von Helmholtz was the son of August Ferdinand Julius Helmholtz who matriculated in 1811 in the Theological Faculty in Berlin. After the war of 1813-14 against Napoleon, in which he participated, he gave up his theological studies, and transferred to the study of classical languages. Following a period of private tutoring he was appointed to the post of form-master at the Potsdam Gymnasium in Berlin in 1820. In this post he devoted himself to becoming one of the most distinguished teachers of the Gymnasium, and received frequent ovations from his pupils. His teaching duties included German, Latin, Greek, of which he was particularly fond, and even mathematics and physics. In addition to these he found time for painting, in which he was self-taught, and for philosophical study.

After Ferdinand's appointment to the Gymnasium he married Caroline Penn, the daughter of a Hanoverian artillery officer, who was a descendant of William Penn, the founder of Pennsylvania. Their first child Hermann, born in 1821, was baptised in the Lutheran Church. For the first seven years of his life Hermann was not physically strong, but had the advantage of having access to wooden blocks for play, from which he derived a useful knowledge of three-dimensional geometry.

His father's influence in poetry, art and music was coupled with the importance of all his children becoming good patriots. In addition he instilled in them a love of both German and Greek poetry, particularly Homer. But by the time Hermann had reached secondary school he had developed a particular interest in Physics, carrying out experiments with spectacle glasses and a small botanical lens which belonged to his father. It is recorded that while his class were reading Cicero or Virgil, he would be working under the desk tracing the rays of light passing through a telescope. This undoubtedly helped many years later in the design of his future invention of the ophthalmoscope.

When it came to choosing a subject for study at a University, Ferdinand with four children to educate, knew that he would be unable to pay for Hermann's continued study of physics. But bearing in mind the possibility of continued warfare on the continent, the German government had a system of scholarships for training doctors. This guaranteed a complete five year's course of study in Berlin, in return for eight years' consecutive service as an army surgeon.

So Hermann began this course in 1838. It involved forty-eight lectures in the week initially. After two years he passed the anatomical examination and was free to begin the independent scientific work he longed for. In this he worked in a small group headed by Professor Muller the physiologist. "Whoever," said Helmholtz later, "comes into contact with men of the first rank has an altered scale of values in life. Such intellectual contact is the most interesting event that life can offer." And Muller was in this category.

Helmholtz himself was able to contribute to the group through his relative wealth of mathematical knowledge. This had been achieved through private study of the great mathematicians Laplace, Biot and Daniel Bernoulli. The group aimed at founding physiology upon the techniques of physics and chemistry, and Helmholtz's initial papers on animal heat and muscle contraction reflect this approach.

In November 1842 Helmholtz was awarded his M.D. degree for his thesis entitled *The Structure of the Nervous System in Invertebrates*. He was then appointed surgeon to the regiment at Potsdam. Continuing his scientific studies, Helmholtz read to the Berlin Physical Society in 1847 his famous paper "On the Conservation of Energy".

The Chair of Physiology at Konigsberg became vacant in 1848, and Helmholtz was offered and accepted the lower grade of Associate Professor, at the remarkably low age of twenty-seven. By virtue of this appointment he was released from his military duty and commitment to serve a full eight years as an army surgeon. But before leaving Potsdam for Konigsberg he married Olga Von Velten in 1849. Then in 1851 he invented the ophthalmoscope for measuring the radii of curvature of the crystalline lens for near and far vision. In 1853 he made the first of a number of visits to

Britain where he formed a lasting friendship with William Thomson. Then in 1855, partly because of his wife's poor health, and despite the fame which had come to him through his invention of the ophthalmoscope, he decided to transfer in 1855 to the vacant Chair of Anatomy and Physiology at Bonn.

But at Bonn he was not happy for long. His research had been in physiology, and the equipment available for research in Bonn was less than in Konigsberg. His lectures in anatomy were not, initially, as good as he would have wished either. Nevertheless his tireless labour enabled him to produce the first volume of his *Handbook of Physiological Optics* in the year after his arrival there. And in the same year he reported to William Thomson that he had taken up the study of physiological acoustics. In this he had discovered that if m and n are the frequencies if two simultaneously sounding tones, that another tone of $(m+n)$ beats is produced as well as the well-known $(m-n)$ beats.

Moreover, in Bonn he contributed one of the series of popular lectures for which he has become famous. This was entitled "On the Physiological Causes of Harmony in Music", delivered appropriately in the native town of Beethoven, one of the heroes of harmony. The concluding paragraph of this essay may be worth repeating:

"The phenomena of agreeableness of tone, as determined solely by the senses, are of course merely the first step towards the beautiful in music. For the attainment of that higher beauty which appeals to the intellect, harmony and disharmony are only means, though essential and powerful means. In disharmony the auditory nerve feels hurt by the beats of incompatible tones. It longs for the pure efflux of the tones into harmony. It hastens towards the harmony for satisfaction and rest. Thus both harmony and disharmony alternately urge and moderate the flow of tones, while the mind sees in their immaterial motion an image of its own perpetually streaming thoughts and moods. Just as in the rolling ocean, this movement, rhythmically repeated, and yet ever varying, rivets our attention and hurries us along. But whereas in the sea, blind physical forces are at work, and hence the final impression on the spectator's mind is nothing but solitude—in a musical work of art the movement follows the outflow of the artist's own emotions. Now gently gliding, now gracefully leaping, now violently stirred, penetrated or laboriously contending with the natural expression of passion, the stream of sound in primitive vivacity, bears over into the hearer's soul unimagined moods which the artist has overheard from his own, and finally raises him up to that repose of everlasting beauty, of which God has allowed but few of His elect favourites to be the heralds."

Helmholtz's scientific reputation had become such that he was given an offer of a Chair in Physiology alone at Heidelberg. When the authorities in Bonn heard this they offered both an increase in salary and the reconstruction of the Anatomy building for his work in Bonn. But some time after this the decision on the reconstruction was postponed, and Helmholtz, after taking his father's advice, moved to Heidelberg in 1858. It has been said by Koenigsberger, the biographer of Helmholtz, that "in Heidelberg with Bunsen and Kirchhoff an era of brilliancy was inaugurated such as seldom existed for any university, and will not readily be seen again."

But the year after the transfer to Heidelberg saw the death of Olga, Helmholtz's wife, who had been ill for a considerable time. There were two children of the marriage, and a family arrangement was made for their immediate upbringing. In a little over a year afterwards Helmholtz remarried. His wife this time was Anna von Mohl, the daughter of a Heisenberg Professor. She had lived in Paris for some time and was fluent in both French and English. When Helmholtz wrote about the short time interval following Anna's death he made the comment, "When love has obtained permission to germinate, it grows without further approach (recourse) to reason."

At the Festival of the Bavarian Academy held in March 1859, Helmholtz met the King of Bavaria who expressed the hope that his future discoveries in acoustics would benefit the architecture of public buildings. But Helmholtz held out small hopes of this, and it may be that these hopes are still not large.

In 1861 Helmholtz visited Britain to lecture on "The Physiological Theory of Music". While there he met William Thomson the Irish-Scot, with whom he developed a close relationship. This lasted over fifty years. As an example of their friendship Helmholtz told Thomson of Kirchhoff's discovery of metals in the solar spectrum. The double dark line D in the solar spectrum proved there is sodium vapour in the sun's atmosphere. But there was also admiration between Helmholtz and Kirchhoff, who was able to say,

"I am content if I can even understand a single word of Helmholtz, but there are still many points in his great book on Acoustics that I cannot unravel."

In 1862 Helmholtz published his work *Sensations of Tone*. This has been described as "the *Principia* of Physiological Acoustics". He showed that the quality of tone depends on the number and intensity of the overtones contained within it. And he produced the fixed pitch theory of vowel tone formation, in which he showed that the pitch of a vowel depends on the resonance of the mouth, independently of the pitch of the note. He also pointed out the extraordinary development of memory in musicians, who are able to execute a formidable number of compositions without having any notes in front of them.

In addition to publishing original research Helmholtz acted as co-translator of Thomson and Tait's *Textbook of Theoretical Physics*, a translation which came out in 1871. Helmholtz's friendship with Thomson has already been mentioned, but his relationship with Tait at Edinburgh is also of interest. "Mr Tait," he says, "thinks of nothing here beyond golfing. I had to go out too; my first strokes came off—after that I hit either the ground or the air. Tait is a peculiar sort of savage, living here, as he says, only for his muscles, and it was not till today, on the Sabbath, when he might not golf, and did not go to Kirk either, that he could be induced to talk of reasonable matters." But in the year after the translation appeared a book was published by Friedrich Zollner, a former student of Weber's, entitled *On the Nature of Comets*. In the Preface to this book Zollner attacked Thomson and Tait's remark that Weber's atoms and forces are not only useless but even harmful. Zollner believed that the mind, in harmony with nature, could discover nature's workings, and that the laboratory could not add to this. A further attack on Helmholtz by Zollner was made after Helmholtz translated Tyndall's works into German. It appears that the reason for this was that Zollner had been converted to spiritualism. His hatred was thus directed in the first place against Tyndall who had embarked on a vigorous campaign in England against spiritualism, and then against Helmholtz who had translated Tyndall's works. These attacks caused considerable pain to Helmholtz.

With regard to possessing gifts of inspiration, Helmholtz said, "So far as my experience goes a flash of inspiration never comes to a wearied brain, or at the writing table." He also quoted a remark of Gauss, who said, "The law of induction was discovered on January 23, 1835 at 7 a.m. before rising." An additional relevant remark Helmholtz made was, "The least trace of alcohol is sufficient to banish inspiration."

Helmholtz never worked out the details of his lectures, but composed them as he went along. He spoke slowly, deliberately, and at times a little haltingly. His eyes looked away beyond the audience as though he were seeking a solution of a problem at an infinite distance.

But when he obtained the knowledge he was seeking, it never seemed to him that this should be the sole aim of mankind. For him it was in action alone, that he found a worthy destiny. Helmholtz never courted extremes in religion matters. By education and conviction he was religious in the noblest sense

After a visit to Oxford in 1863 Helmholtz remarked, "I now understand the devotion of the Englishman to his university. Their system works wonderfully well for the education of a gentleman, but it cannot lead to much in science, and it needs an extraordinary interest in science to prevent a Fellow from sinking into indolence."

Helmholtz's stay in Heidelberg lasted until the death of Professor Magnus in Berlin in 1871. The department of Physics in Berlin was the star department in this field in Germany. Both Helmholtz and Kirchhoff were considered for the post of successor to Magnus, and Kirchhoff was offered the appointment, but turned it down. The reasons given by the University of Berlin in their recommendation to the appointing Minister were:

> "If Helmholtz is the more gifted and universal in research, Kirchhoff is the more practical physicist and successful teacher. While Helmholtz is the more productive, and is always occupied with new problems, Kirchhoff has more inclination to teaching; his lectures are a pattern of lucidity and finish; also from what we hear he is better able to superintend the work of elementary students than Helmholtz."

When this offer was made to Helmholtz he stipulated the following conditions had to be met before he would accept the post:

(1) A personal salary of £600.
(2) A Physical Institute shall be built with the necessary equipment for instruction for the private work of the Director, and for the practical work of the students.

---

(3) The promise that he would have sole charge of this Institute, and of the collection of Instruments. The Auditorium in the Physical Institute must equally be retained for my sole use.

(4) An official lodging for myself in the Institute, and a corresponding allowance for rent until it shall be ready.

(5) Provisional use of rooms hired in the vicinity of the University for my own work in Physics and for some of my students.

(6) A proper allowance for the expense of moving.

The Minister of Education lost no time in applying for the necessary funds to the Minister of Finance, and all the funds were made available through fund transfers from the General Revenue. The stipend was to be paid during the course of his life. Had it not been for the promptness of this official response it is possible that Helmholtz might have gone to Cambridge as Professor of Experimental Physics, because a few days after his return from Berlin he received a letter from Sir William Thomson asking if he were disposed to accept this Professorship.

Towards the end of Helmholtz's period in Heidelberg his interests had turned towards electrodynamics, and in 1871, the year he took up his Chair in Berlin, he announced that the velocity of electromagnetic induction was greater than 314,000 miles per second. But his greater aim was to bring order to a confusing field, and to do this through the presentation of theory rather than experiment. As pointed out by J. D. Buchwald in *The Creation of Scientific Efects* there were two competing forms of electromagnetism current at the time:

(A) The Neumann-Helmholtz Approach

This listed the following principles:

(1) Electrically charged bodies interact through the Cavendish force which can be obtained through the gradient of a Potential V1 that depends solely on the distance between the charges and their magnitudes.

(2) Current bearing circuits interact with one another by means of a mechanical force which can be obtained through the gradient of a

potential function V2 that depends on the distances and the orientations of the current elements and on the intensities of the currents.

(3) Current bearing circuits also interact with one another through an electromotive force that is given exactly by the time derivative of V2.

(4) The current in a circuit is proportional the net force that drives it.

(5) The charge density p at a point will change only if the current there is inhomogeneous, according to the equation of continuity that the divergence of the current equals minus the rate of change of the charge density.

(6) All interactions in electrodynamics require the existence of a corresponding system of energy.

The second competing form of electromagnetism at the time was:

(B) The Fechner-Weber Approach.

This listed the following principles:

(1) Charge consists of two kinds of electric particles, or atoms of electricity.

(2) The electric current is the equal and opposite flow of these particles.

(3) The particles exert central forces on each other that depend upon their distances, as well as on the first and second time derivatives of these distance. Consequently all electromotive forces—mechanical and electromotive—drive from a single, fundamental action.

(4) The current in a circuit is proportional to the net force that drives the electric particles.

The Fechner-Weber approach has the pedagogic advantage of focussing on the physical structure of the current. Helmholtz himself later subscribed to this view, as is seen in his 1881 Faraday lecture to the Fellows of the Chemical Society. This lecture has been described by Lord Kelvin as "an epoch-making monument of the progress of Natural Philosophy in the nineteenth century, in virtue of the declaration, then first made, that electricity consists of atoms. Before that time atomic theories of electricity had been noticed and rejected by Faraday and Maxwell, and probably by many other philosophers and workers; but certainly accepted by none."

In addition to asserting that electricity consisted of atoms, Helmholtz obtained an expression for the vector potential of a single current element, which in all cases where the current is closed gave the same value as Neumann's and other early workers' formulas. But where the current circuit is open, his equation of continuity differed from that of Maxwell. In Maxwell's analysis a changing electric field is itself a current, but for Neumann and Helmholtz this is not so. His answer for the new potential differs from that of Neumann, Maxwell, and Weber containing a constant k which is equal to 1 for Neumann, zero for Maxwell and -1 for Weber. Helmholtz then went on to show that negative values were unstable, and it is only when k is equal to zero that the waves of propagation can be wholly transverse. For k equal to 1 the movement of electricity differs little from that with k equal to zero, but allows a longitudinal polarisation which is not observed in practice.

Thus Maxwell's theory came to be accepted throughout Germany and the continent of Europe in the 1880's, almost entirely due to the thinking of Helmholtz. In 1881 he was given the degree of Doctor of Laws by Cambridge University. In 1883 he was ennobled by William I. At the 300 hundredth anniversary, in 1884, of the founding of Edinburgh University, he represented the Berlin Academy, and in 1890 he did likewise for the six hundredth anniversary of the University of Montpelier. In 1891 he was given the title of Excellency by William II, and in 1899 Einstein said of him, "I admire ever more the original free thinker, Helmholtz."

And in addition he accepted in 1878, Heinrich Hertz as a research worker in Berlin, about whom more will be said later in this chapter.

William Thomson (Lord Kelvin) (1824-1907)

Since William Thomson was much influenced by his father James Thomson, it is helpful to begin this study by describing the father's background. James Thomson was born at Annaghmore Farm, about ten miles south of Belfast. He was brought up as a farm labourer, but received from his father the rudiments of education. In addition he studied for himself, so that he was able to construct a sun-dial when he was eleven or twelve years old. He then advanced his knowledge by making a night-dial to tell the time by the position of one of the stars in Ursa Major.

In view of William's intellectual abilities, his father allowed him to go to a small school kept by the Minister of the Secession Presbyterian Church nearby, to learn classics and mathematics. Soon he was promoted to be Assistant Teacher. At that time it was possible to study in Glasgow University from 1st November to 30th April, and keep his teaching job as an Assistant Teacher, by working there from 1st May to 31st October. After four years study at Glasgow University he graduated M.A. in 1812. In 1814 he was appointed a Teacher of Mathematics at the Royal Belfast Academy, and in the following year he was made Professor of Mathematics in Belfast.

James Thomson married Margaret Gardner, the daughter of a Glasgow merchant, in 1817, and there were seven children of the marriage. The two older boys, James, born in 1822 and William born in 1824, both became Professors. James became Professor in Engineering at Belfast, and later in Glasgow, and William in Natural Philosophy in Glasgow.

In 1829 James was given the honorary degree of Doctor of Laws by Glasgow, which was a cause of rejoicing among the family. But in the following year his wife Margaret died when William was aged six. James then took on the responsibility of teaching the boys himself, apart from sending them to a writing school in Belfast. In particular he taught them how to use globes, and he introduced them to Latin.

James Thomson was then appointed to the Chair of Mathematics in Glasgow. He kept the education of his sons in his own hands, but in addition wrote mathematical textbooks of a high quality. It appears that James (junior) and William were allowed to attend some of their father's lectures unofficially. They also constructed electrical machines, Leyden jars and batteries.

In 1834 James and William matriculated at Glasgow University, when James was twelve and William was ten. At the end of the session, William obtained two prizes in the Humanity (i.e. Latin) class before he was eleven. In the following session he obtained prizes in Natural History and Greek, and in the following year both brothers obtained prizes in Junior Mathematics. Proceeding to the Senior Mathematics Class they again stood at the top, and in addition William received the Second prize in Logic. In the following year they took first and second prizes in Natural Philosophy, and in the year after that William gained the class prize in Astronomy. Their final year at

University from 1840-1841 saw William achieve fifth place in the Senior Humanity Class. But a much more important event occurred in that session when John Nichol, the Professor of Astronomy, brought to his notice *Theorie analytique de la chaleur*, by Fourier. This had been published in Paris in 1822, and although Nichol did not profess to have really read it, he was capable of perceiving its greatness, and making William appreciate it.

So William is on record as saying that on 1st May 1840 he took Fourier out of the University Library, and in a fortnight had mastered it. It is fair to put on record that John Nichol's predecessor, Professor Meikleham, had previously taught his students reverence for the great French mathematicians, Legendre, Lagrange and Laplace. And its refinement of style was such as to cause Clerk Maxwell to pronounce it "a great mathematical poem". The knowledge of French in possession of the boys had been increased during the previous summer when they were taken to Paris for two months. After that experience their father had taken them to Germany to provide them with an equal opportunity of learning German. But in William's case he had become side-tracked by a book on *Heat* by Kelland of Edinburgh which purported to say that Fourier's book was wrong. In fact Thomson discovered the reason for Kelland's error, and this provided him with the opportunity of writing his first research paper. Professor Kelland had been appointed to the Chair at Edinburgh University after being Senior Wrangler at Cambridge. He was an algebraist rather than a geometer, and on this ground alone Sir W. Hamilton, who held the chair of Logic and Metaphysics in Edinburgh, had argued that the appointment should have gone to Gregory who had been Fifth Wrangler in Cambridge, and was a geometer. Sir W. Hamilton's argument was as follows:

"The mathematical process in the algebraic method is like running a railroad through a tunnelled mountain; that in the geometric method is like crossing the mountain on foot. The former, the algebraic method, carries us by a short and easy travel, to our desired point, but in miasma, darkness and torpidity, whereas the geometric method allows us to reach it only after time and trouble, but feasting us at each turn at glances at the earth and of the heavens, while we inhale health in the pleasant breeze, and gather new strength at every effort we put forth."

Despite this argument the Town Council appointed Kelland, albeit by a narrow majority.

April 1841 brought the end of William's student days in Glasgow. He left without taking a degree, apparently on the grounds that it might prejudice his chances of being accepted by Cambridge as an undergraduate.

William Thomson at Cambridge University

Wm. Thomson became resident in Cambridge in October 1841. He entered St Peter's College (now Peterhouse), probably because his father knew of the tutoring skills of Hopkins, their mathematical coach. It is also true that it is the oldest College in Cambridge, having been founded in 1257 and given a Charter in 1284. It is also the smallest of the Colleges, with at present 284 undergraduates, 130 postgraduates and 45 Fellows. There also appear to be Scottish connections with the College. Both Tait and Clerk Maxwell who will be considered later in this chapter were students there. And it has an engineering reputation since Charles Babbage of first mechanical computer fame went there. Likewise so did Christopher Cockerell, the inventor of the Hovercraft, and Frank Whittle who invented the jet engine.

In 1842 Wm. Thomson obtained Hopkins as his private tutor and this arrangement lasted for the remainder of his time at Peterhouse. There were, of course, other excellent coaches in other Colleges in Cambridge, each with his own characteristics. Wm. Thomson read for eight hours each day, but also walked, boated and rode. In his first year the total cost of his maintenance has been given as £220. In one of his letters home he drew attention to the fact that he had 15 yards of bookshelves, but only 0.5 yards of books.

Foundation of Cambridge Musical Society

This originated at an informal private amateur concert at Peterhouse. When on a later occasion a larger room was required, an arrangement was made to hire the local Red Lion Hotel. But this required the permission of the Master of Peterhouse, who refused to give it unless the Peterhouse Society changed its name to the Cambridge Musical Society. The first performance was given on 8th December 1843 with eleven instrumentalists, with Wm. Thomson playing the French Horn. But the founder of the University Musical Society was G. E. Smith who entered Peterhouse as a Freshman at the same time as Wm. Thomson in 1841, but who died in 1844. It has been reported that the strongest player was William Blow on the violin, but the Society was also helped by a very good tenor singer, C. M. Ingleby. The Society had quartet evenings as well, with one member approving of nothing more modern than Gregorian music.

Second Wrangler and Smith's Prizeman

Wm. Thomson profited from his Cambridge training. But it was a system in which examination results depended much on getting through a large amount of bookwork in a short time. Two days of examination depended on this textbook work rather than on problems requiring analytical examination. The successful Senior Wrangler of this year, by the name of Parkinson, it was said had practised writing out against time for six months together, merely to gain pace, and the "Pace of Parkinson" has become almost a proverb in Cambridge.

But for the Smith's Prize different qualities were looked for. The examination is of a higher character than that of the Senate House which determines who shall be Senior Wrangler. It is intended to furnish a higher test of the merits of the first men. And it was later made known that the four examiners were unanimous in their verdict. Wm. Thomson had beaten all his competitors in all the papers. In two of them the marks were in the proportion of three to two. The Master of Peterhouse also said that in this examination he had proved himself decidedly superior to the Senior Wrangler.

The Age of the Earth

It has previously been stated that Wm. Thomson wrote his first research paper based on his admiration for Fourier's *Theorie analytique de la chaleur*, when it was challenged by Kelland. During his first summer vacation in Cambridge he continued his study of heat. He produced a solution for the linear motion of heat in an infinite solid, starting from a specified value of temperature in a given zero-plane. He thus obtained the temperature at any distance x from a given zero-plane at any time "t" thereafter. Naturally he then considered what information could be gained by considering negative values of time. His conclusion was that no such information was possible, and this convinced him that there must have been a beginning to the cosmos.

Postgraduate Study in Paris and Peterhouse

In January 1845 Wm. Thomson went to Paris with a copy of George Green's essay, "An Essay on the Application of Mathematical Analysis to the Theories of Electricity and Magnetism." This had been printed in Nottingham in 1828 by private subscription, and it appears that fewer than a hundred copies were printed. He had not been able to get a copy in the Cambridge Library or in any bookshop, but found that Hopkins, his previous coach, possessed three copies. Two of these Hopkins handed to him, one for himself and one for Liouville, the editor of the periodical *Journal de mathematiques*, whom Wm Thomson was expecting to meet in Paris. In fact, it was in this journal that Wm. Thomson's famous papers on electric images were first published, and he was responsible for having introduced Green to French physicists.

Among other mathematicians for whom Wm. Thomson was given introductions was A. L. Cauchy (1789-1857), the author of 789 research papers. He seemed to produce enemies among his colleagues, and was a strong supporter of the Jesuits. Wm. Thomson believed, along with others, that Cauchy tried to convert him to Roman Catholicism. But Kelland, who was an Anglican clergyman, in writing to Cauchy, referring to his own wife's recent death, expressed his hope "that she and they will meet in heaven". So the hostility against Cauchy was not universal.

The reason for the visit to Paris, which lasted for more than four months, was that the French had become experts in experimental techniques. At the College de France, where Professor Regnault was the professor of Natural Philosophy, the Government had given a great deal of money for apparatus for popular experiments, and for historical illustrations of the lectures. In Professor Regnault they had a man who had conducted a marvellous series of researches on steam engines, and was a master of the art of minute and accurate experiment. At that time there was no provision for teaching experimental physics in French Universities, so the techniques had to be learned in research laboratories. Wm. Thomson sometimes went to the laboratory at 8 a.m. and stayed till 5.p.m.

In addition to having introduced George Green to French physicists, Wm. Thomson was also responsible for having rehabilitated Benuit Paul Emile Clapeyron to French scientists. Clapeyron's paper "Memoir on the motive power of fire" expounded in a more useful form the importance of the Carnot cycle, expressed as a closed curve on a chart of pressure against volume. But Sadi Carnot's original paper, like Green's paper, was difficult to obtain. In 1845 Wm. Thomson searched for it in vain in Paris bookshops, and did not obtain a copy of it until 1848. This was from Professor Lewis Gordon, the first Professor of Engineering at the University of Glasgow.

Wm. Thomson returned to Cambridge after spending four and a half months in Paris. It was then necessary for him, financially, to take on pupils, coaching them in Mathematics. But in January 1845 he was appointed Foundation Fellow, which was worth £200 p.a. along with rooms at the College. He held this until 1852 when he vacated it by virtue of his marriage. Then in 1872 he was re-elected as a Life Fellow on the grounds of his eminence, notwithstanding his marriage.

Appointment to the Glasgow Chair of Natural Philosophy

In May 1846 Professor Meikleham, who held the Chair of Natural Philosophy in Glasgow University, died. Wm. Thomson's father, James Thomson, had been grooming his son to have the necessary qualifications and experience so as to maximise his chances of taking over this chair. Testimonials were obtained from twenty-eight persons, and there were six applicants for the post. It is of interest that Michael Faraday made it a rule never to give a testimonial, but he wished Wm. Thomson success.

The Faculty in Glasgow appointed Wm. Thomson to the vacant chair after he had submitted a Latin Essay on "The movement of heat through the body of the earth". The dissertation totalled about one hundred and fifty words only. It was also required that he took the oaths to government, and that he further promised on the first convenient opportunity, to subscribe the formula of the Church of Scotland as required by law.

The Young Professor

Wm. Thomson was twenty-two years of age on his appointment to the Glasgow Chair. Since the university year extended from 1st November to 30th May only, this left him much freedom for the summer months. In the first few years of his professoriate he went to Cambridge to make new acquaintances like Maxwell or Tait, or renew acquaintance with Peterhouse friends. He also rowed in an eight-oar boat, played in the University Musical Society as second horn, and took part in foreign tours. Then in the autumn he had to be back in Scotland, with obligations to his family and to boat on the Clyde.

In 1848 he visited Stockholm which he described as the most beautiful city he had ever seen. Two years later he was in Paris, and on this visit he records that he attended the English Church where the sacrament was administered to a very large congregation. The following year, 1851, saw him elected F.R.S., and one of the signatures on the Certificate of Candidature was that of Michael Faraday. In that same year he spent two months away from England, visiting Regnault in Paris, and afterwards writing out for him his own theory of the dynamics of heat.

Marriage was the most important event of his life in 1852, when his wedding took place to Margaret Crum, whom he had known since boyhood. She was aged 22 while he was now 28. It is recorded of her that she was well read, possessed of a lively imagination and a poetic fancy, but was also of a deeply religious nature. During her married life she composed a number of poems, including translations from German poets, but she died at the age of forty. It was necessary for Wm. Thomson to give up his fellowship at Peterhouse, since such fellowships were restricted to bachelors.

In the same year as the marriage Wm. Thomson undertook a study of oscillatory discharges in circuits which possessed resistance, capacitance and inductance. It was known that discharges of lightning sometimes occurred in multiple flashes rather than a single flash. So Wm. Thomson analysed the problem from energy considerations and discovered that a critical relation existed if the capacity in the circuit was equal to four times the inductance divided by the square of the resistance. If the capacity was less than this, the discharge was oscillatory. If the capacity was greater than this, the discharge was non-oscillatory, the charge dying away without reversing. He then went on to suggest that it might be possible, by discharging a Leyden Jar, or another small capacity, through a circuit having large inductance and small resistance to produce artificially such oscillatory discharges. Such oscillatory sparks were photographed in 1859 by Fedderssen. The idea of these discharges is now the basis of radio communication.

Wm. Thomson and his Collaborators

Wm. Thomson, up to the time of his appointment to the Glasgow Chair, would have been a man who worked on his own, establishing his own scientific reputation. But immediately prior to this appointment, after his return from Paris in 1855, he met George Stokes, at that time a fellow of Pembroke College in Cambridge, who had been Senior Wrangler in 1841, the year of Wm. Thomson's entry to Peterhouse. George Stokes was an individual worker, making his own experiments without the assistance of students. He was methodical and cautious, while Wm. Thomson was stimulating and speculative. For more than fifty years they communicated to each other the progress of their ideas. Thus the bright lines in the spectrum of sodium, and the reversal of them giving a black line of absorption in place of the characteristic yellow, they saw was explained by the vibrating particles taking up the energy of the lines attempting to pass through.

Another collaborator whom Wm. Thomson met in 1855 also, was Hermann Von Helmholtz. When Mrs Thomson was in Germany, taking the waters of a Spa, the opportunity arose of Wm. Thomson meeting Von Helmholtz for the first time. This meeting was described to Mrs Helmholtz by her husband as follows:

> "I expected to find the man, who is one of the first mathematical physicists of Europe, somewhat older than myself, and was not a little astonished when a very juvenile and exceedingly fair youth, who looked quite girlish, came forward.—He far exceeds, in intelligence and lucidity, and mobility of thought, all the great men of science with whom I have made personal acquaintance, so that I felt quite wooden beside him sometimes."

Wm. Thomson and his Royal Institution Lecture 1856

As an indication of Wm. Thomson's breadth of interests, the following peroration from his lecture at the Royal Institution delivered on 29th February 1856 is given:

> "The opening of a bud, the growth of a leaf, the astonishing development of beauty in a flower, involve physical operations which completed chemical science would leave as far beyond our comprehension as now the difference between lead and iron, between water and carbonic acid, and between gravitation and magnetism, are at present. A tree contains more mystery of creative power than the sun, from which all its creative energy is borrowed. An earth without life, a sun, and countless stars, contain less wonder than that grain of mignonette."

This breadth of interest was apparent to others as well as to scientists. When Wm. Thackeray was in Glasgow he dined twice with Wm. Thomson. A common acquaintance was Dr John Brown, the physician and essayist who said, "Thackeray was delighted with your Wm. Thomson; he said he was an angel and better, and must have wings under his flannel waistcoat. I said he had, for I had seen them!"

The Atlantic Telegraph

By the year 1850 overland telegraphy had become a prosperous business, and attention naturally turned to increasing the financial profits by extending the techniques to underwater telegraphy. In 1851 the Dover-Calais line had been successfully laid, and this was followed by others linking England with Ireland and Holland. But a line across the Atlantic would be two thousand miles long, and in places the ocean depth would be as much as three miles. Another problem arose with regard to signalling speed. A signal which is sent as a short, sudden impulse is changed in character, for long cables, by being smoothed out into a longer lasting impulse, which rises gradually to a maximum and then dies away. Wm. Thomson showed theoretically that the delay is proportional both to the capacity and the resistance of the cable. Since each of these is proportional to the cable length, it means that the delay is proportional to the square of that length. Thus if a cable 200 miles long showed a delay of 0.1 second, one 2000 miles long would have a delay of 10 seconds. To overcome this problem the diameter of the central conductor can be increased to reduce the resistance, and by increasing the thickness of the gutta-percha insulation the capacity can also be reduced. Thus the delay of a long cable can be kept the same as that of a shorter cable. This argument was not universally accepted unfortunately.

The manufacturers completed their work in making the cable in 1857. The size of this cable was not in accordance with the recommendation of Wm. Thomson, nor was an opportunity given to him to test the cable before laying. After a number of mishaps, which included having to re-lay some of the cable, a message was transmitted in August 1858 from the Queen to the President of the United States containing 99 words. This took sixteen and a half hours to transmit. But the same message was transmitted back from Newfoundland in sixty-seven minutes. Clearly there was something seriously wrong. Matters gradually worsened and the last successful transmission was obtained in October 1858. The trouble was associated with mechanical damage to the cable, possibly produced during storms in the Atlantic.

In 1860 the President of the Institution of Civil Engineers reported that upwards of 9000 miles of submarine telegraph cable had been laid, but not more than 3000 miles of it could be said to be in working order. But £600,000 of new funding became available. In addition an artificial cable had been devised to study in the laboratory the properties of cables in general, and from this study a signal sharpening condenser had been incorporated to increase the working speed. A scientific committee had been formed. This included Wm. Thomson, and a decision was made to choose a copper conductor three times as thick as in the earlier cable.

In 1865 the cable was shipped at Greenwich into the *Great Eastern*, which was lying idle. This ship was equipped with both screw and paddle propulsion so was well equipped for manoeuvring. Messages were sent through the coiled up cable at 3.8 words per minute. The staff and crew numbered about 500 persons. On July 23rd 1865 the *Great Eastern* started westward, and 1200 miles had been laid when all signals ceased. The broken cable lay in 2100 fathoms and a decision was made to wait another year. Capital for 1800 miles of cable was raised so that the *Great Eastern* could sail again in May 1866. Wm. Thomson had to go to London frequently to test the cable during manufacture, and his secretary used to be sent to the

Glasgow Railway Station a few minutes before the mail train left with the urgent message from Thomson: "I have gone to White's to hurry on an instrument. The London train must on no account start tonight until I come." And such was the national importance of the Project, and such the honour in which Wm. Thomson was held, that the Station Master obeyed.

On 13th July 1866 the *Great Eastern* started westward from Ireland and arrived at Heart's Content Bay on 27th July 1866. Within twenty four hours the line was busy with messages from Europe. The broken cable from 1865 was then lifted, so that two cables were available by September 1866.

Honours for Wm. Thomson poured in though some more slowly than others. In September 1866 he was knighted. In November 1866 he was given the Freedom of Glasgow. In April 1869 he was given the LL.D by Edinburgh University. In 1870 he was offered the Directorship of the new Cavendish Laboratory in Cambridge, but turned it down, as he did again later.

James Clerk Maxwell (1831-1879)

James Clerk Maxwell's paternal grandfather was a Captain James Clerk, who worked as a sea captain with the East India Company. There is a family story that on one occasion his ship was wrecked on the coast of India. Captain Clerk swam to land using the bag of his bagpipes as a float; then climbing ashore he began to play the pipes uncommonly loud, thereby cheering the survivors who were still swimming, and, in addition keeping the Bengal tigers at bay.

His father was John Clerk Maxwell, one of the Clerks of Penicuik, and his mother was Frances Cay of Northumberland. Their daughter Elizabeth had died in infancy and James was their only son. John Clerk Maxwell was the laird of the estate of Middlebie. This estate did not have a dwelling house for the laird, so John Clerk Maxwell lived in Edinburgh. After he finished University he trained as an advocate, but he was also a Fellow of the Royal Society of Edinburgh, and published one scientific paper, a proposal for an automatic feed printing press. After his mother's death he married and moved to 14 India Street which was built for them, and it was here that their son James was born.

James's mother played the organ and composed some music, and knitted well. She died at age 48, after an unsuccessful operation without the benefit of a general anaesthetic. John Clerk Maxwell was fifty-two at the time and did not re-marry.

At the time of James's birth his parents had been living for some years in Glenlair, the house designed by, and built under the supervision of, John Clerk Maxwell himself. But they moved to India Street for the birth, to the house which they had retained. So James grew up in Glenlair. At the foot of the garden a place was hollowed out in the bed of the burn, and this was used for bathing.

When James was two and a half he showed an interest in doors, locks and keys, and in the wiring and sounding of bells within Glenlair. At the same age he was given a tin plate to play with, and discovered with it how to reflect and direct rays from the sun in chosen directions. The influence of his father meant that sections of a globe cut out with the different constellations, were available for him to ponder.

James's mother had charge of his education until her last illness in 1839. She was an Episcopalian, while his father was Presbyterian. With regard to church attendance this meant that James was taken to the Presbyterian Church in the morning, and to the Episcopalian Church in the evening. Consequently he was familiar with both types of service and profited thereby. Because of his excellent memory he became able to recite the whole of the 119th Psalm, and it is said that he was able to identify from which Psalm almost any quotation came. His mother also encouraged him to "look up through Nature to Nature's God". This brought out the sensitive nature with which he was endowed, and allowed him to recite with feeling the words of Robert Burns:

> "O Nature! all thy shows and forms
> To feeling, pensive hearts have charms!
> Whether the summer kindly warms,
> With life and light;
> Or winter howls, in gusty storms,
> The long, dark night!"

And the following verse is particularly applicable to his later manhood:

> "The Muse, no poet ever found her,
> Till by himself he learned to wander,

Adown some trotting burn's meander,
And no think long:
"O sweet to stray, and pensive ponder
A heart-felt song."

In nature he loved living things, particularly frogs and tadpoles. It was one of his favourite games to jump like a frog, and a pastime to put a young frog in his mouth, and let him jump out again.

After the death of his mother, when he was aged eight, a young tutor aged sixteen was engaged for two years. This period appears to have seen a continuous struggle between the tutor and his pupil. James Clerk Maxwell's habit of giving oblique answers is reported to have started at this period in his life, when the answers he gave to the tutor appeared not to answer the questions asked, but did. This understandably led to friction between them, an example of which has been recorded in a sketch by a family member. This shows the tutor attempting to reach with a long handled rake the tub in which the pupil has taken refuge from him on the duck pond at Glenlair, with the other members of the family looking on.

At the age of ten, James Clerk Maxwell was sent to school at Edinburgh Academy, founded by, among others, the novelist and poet, Sir Walter Scott. The stated objective of this school was to provide Scottish youth with a classical education along English lines. Among his fellow pupils was Lewis Campbell, a nephew of Thomas Campbell the poet. This Lewis Campbell later became Professor of Greek at St Andrews, and Clerk Maxwell's biographer. Another pupil was P. G. Tait who was Senior Wrangler and Smith's Prizeman in Cambridge in 1852, and Professor of Mathematics in Belfast, where he made the acquaintance of Sir William Hamilton the inventor of quaternions, in 1853. Later he became Professor of Natural Philosophy in Edinburgh from 1860. Tait was described by Hermann Von Helmholtz "as a peculiar sort of savage, living in Edinburgh for his muscles, and it was only on the Sabbath, when he might not play golf, and did not go to kirk either, that he could be induced to talk of reasonable matters." That was in 1871, but only four years later he published a book entitled *The Unseen Universe*, with Balfour Stewart as co-author, in which his strong religious feelings became apparent.

When James Clerk Maxwell arrived at Edinburgh Academy in the second month of the school's second year, it was to an experience he would not forget. His Galloway accent was different from that of Edinburgh, and more importantly his clothes and shoes had been designed by his father, and were ridiculed by his peers. A contemporary pupil later wrote of him:

> "Clerk Maxwell when he came to the school was somewhat rustic and somewhat eccentric. Boys called him 'Dafty' and used to try to make fun of him. On one occasion I remember he turned with tremendous vigour, with a kind of demonic force, on his tormentors. I think he was left alone after that, and gradually won the respect even of the most thoughtless of his schoolfellows."

James seldom took part in games, choosing to play marbles or doing gymnastics on the few trees in the school grounds. If his father was in Edinburgh they walked together, seeing Leith Fort, the preparation for the Granton railway, and examining the stratification of the Salzburg Crags. In Glenlair his father would try to cheer him up by concocting the wildest absurdities. At school his best subjects were Scripture, Biography and English. And he began to make friends. Two years after his arrival he began what was to be a lifelong friendship with Lewis Campbell, later to be Professor of Classics at St Andrews University, and the one who was to be his chief biographer. It was fortunate that the boys lived at Numbers 27 and 31 Heriot Row respectively, so that contact was easy between their homes.

As James progressed through the school his innate ability came to the fore. In 1845 he received the eleventh prize for scholarship in the school, the prize for English Verse, and the Mathematics Medal. He tried for a prize in Scripture Knowledge, but another in the seventh form got it. But he told his aunt Miss Cay that his friend Lewis Campbell had been awarded six prizes.

About this time James's father became more assiduous in attending both the Edinburgh Society of Arts and the Royal Society of Edinburgh, and he took James with him repeatedly to both. One result was that when James was fifteen years old he devised a mechanical method of drawing oval shapes geometrically. He did this by wrapping a thread round pins such that the multiple distance from one focus plus a multiple distance from the other was a constant. Thus a ratio of 2/3 gives a simple egg shape. This was an extension of the well-known technique for drawing an ellipse where the distance from one focus plus the distance from the other is a constant. The discovery of James's technique was drawn to the attention of Professor Forbes of Edinburgh University by James's father. After the literature was consulted it was agreed that this was a novel result which merited publication by the Royal Society of Edinburgh. But because of James's age the paper was read to the Royal Society by Professor Forbes himself.

This was obviously a critical time in James Clerk Maxwell's life. With a research record behind him it would have been possible for him to have proceeded directly to Cambridge after he finished school in a further two years. But James's father was for several years reluctant to send his son to an unfamiliar distant institution like Cambridge. He was a good man, who was afraid of the religious influences to which his son might be subjected in Cambridge. The beliefs of the Anglo Catholic Pusey, the Anglican divine from Oxford, were reported to be spreading throughout the Anglican Church, and John Clerk Maxwell was afraid that impiety was common in University circles at Cambridge. In addition no decision had at that stage been reached on James' future career. It was still possible at age 17 that he could have become a lawyer in Edinburgh like his father before him. So James spent three years at Edinburgh University.

The Scottish University course required, from around 1750, a two-year study of philosophy, irrespective of the main options selected. It appears that the standard of the Greek and Latin classes at Edinburgh, at least for those who had a good grounding of it at school as Maxwell had, was less high than the standard in Logic, Mathematics and Natural Philosophy. Consequently it was for the study of these three higher standard subjects that Maxwell enrolled for the first two years, Logic and Metaphysics under Sir William Hamilton, Mathematics under Kelland and Natural Philosophy under Forbes. In his third year he studied further Natural Philosophy, Chemistry and Moral Philosophy.

The two Professors who influenced Maxwell most were Sir William Hamilton and Professor Forbes. In the University the Chair of Logic and Metaphysics held by Sir William Hamilton was, by tradition, the most important in the University, and Maxwell himself said that the lectures he received from Sir William were the most solid. These lectures were published later in three volumes, the first two volumes on Metaphysics, and the third on Logic. They carry that stamp of quality in them, which confirm Maxwell's judgment. Philosophy was to Hamilton an end in itself. It was like a building which he was constructing, of which it has been said that the truths of science were the hewn stones used in its construction. It was a way of life to Hamilton, just as Sandemanism was a way of life to Faraday. Although earlier Scottish philosophers had taken the view that mathematics was the nearest thing to perfection which man had devised, Hamilton disagreed. He took the view that mathematics is not adapted to the real problems of life, since it is soluble only when the number of variables is small. But philosophy he saw as central to man's existence, and because of this it should be central to a university education.

It can be readily understood that a Professor as forceful and as scholarly as Hamilton would have a great influence on a certain type of young mind. In Maxwell's case this became most evident in the use he later made of analogies between mechanical and electrical phenomena. Thus Maxwell took kinetic energy in mechanics, as the analogue of magnetic energy stored in an inductive element in electricity, since charged particles in electricity move with a certain mean velocity. Likewise he conceived the idea of potential energy stored by elastic distortion in mechanical engineering as being an analogue of the electric energy stored in the displacement of the same charged particles. Then using his knowledge of the propagation of transverse waves within an elastic substance, Maxwell showed that electromagnetic waves could be propagated in a dielectric medium, with a velocity equal to the velocity of light. Thus the identity of electromagnetic waves with light waves was established.

The staggering boldness of this theoretical prediction was accompanied by an assertion that the mathematical form of Oersted's discovery, given by Ampere in 1821, required to be supplemented by a new type of electric current to make Ampere's Law correct at all frequencies. This additional current then produces a magnetic field even inside a capacitor, where previously only an electric field was believed to exist. Such new predictions and interpretations found difficulty in being accepted, even by a friend of Maxwell's in the person of Kelvin, who appears to have kept on looking for a more mechanical viewpoint. But it is these predictions of Maxwell which have led to the radio, television and radar industries over the last 150 years.

But returning to the study of James as a schoolboy, we learn from his biographer Lewis Campbell that when he visited Glenlair for the first time in 1846, he observed James in church on Sunday sitting preternaturally still, with one hand resting lightly on the other, not moving a muscle, however long the sermon might be. The service was, as it were, photographed on his mind. And Sunday evening closed with a chapter and prayer, which his father read to the assembled household.

In 1847 he entered Edinburgh University to study Mathematics, Natural Philosophy and Logic. During his time there he composed an essay which illustrates the depth of his mind. The subject was "On the Properties of Matter", and the first few lines begin:

"These properties are all relative to the three abstract entities connected with matter, namely space, time and force.

1. Since matter must be in some part of space, and in one part only at a time, it possesses the property of locality or position.
2. But matter has not only position but magnitude; this property is called extension.
3. And since it is not infinite, it must have bounds, and therefore it must possess figure.
   These three properties belong both to matter and to imaginary geometrical figure, and may be called the geometric properties of matter. The following properties do not necessary belong to geometric figures.
4. No part of space can contain at the same time more than one body, or no two bodies can co-exist in the same space; this property is called impenetrability. It was thought by some that the converse of this was true, and that there was no part of space not filled with matter. If there be a vacuum, said they, that is empty space, it must be either a substance or an accident.
   If a substance it must be created or uncreated.
   If created it may be destroyed, while matter remains as it was, and thus length, breadth and thickness would be destroyed while the bodies remain at the same distance.
   If uncreated we are led to an impiety..."

And so it goes on for another two pages, ending with:

"By means of touch, combined with pressure and motion we perceive—

1. Hardness and softness, comprehending elasticity, friability, tenacity, flexibility, rigidity, fluidity, etc.

2. Friction, vibration, weight, motion, and the like.

The sensations of hunger and thirst, fatigue, and many others, have no relation to the properties of bodies."

James Clerk Maxwell undoubtedly benefited from his three years study in Edinburgh. But there were friendship and scholastic advantages in continuing study in Cambridge. Tait who was a year behind him at Edinburgh Academy, had gone directly to Cambridge from that school. Robert Campbell was destined for it, while his brother Lewis was at Oxford. But the chief influence on both James and his father was that of Charles Mackenzie, six years older than James, born in Peeblesshire, and who had attended Edinburgh Academy for a time. In 1844 he had gone to St John's College in Cambridge, but found that as a Scotsman he would be ineligible for a Fellowship there, and so transferred to Caius College. His personality was such that he soon became a favourite at this college, having a remarkable modesty of disposition, and being very good at sports and rowing. As a boy he had learned Hebrew, with the object of fitting himself for the Ministry. In 1848 he had graduated as Second Wrangler, and narrowly missed being the second Smith Prizeman. His elder brother was responsible for the Forbes Mackenzie Act in Parliament which forbad the selling of whisky in Scotland on Sundays and after 10 p.m. on weekdays. Later in life Charles Mackenzie was consecrated Bishop and led the Universities' Mission to Central Africa. He served under David Livingstone but died from malaria at age 37.

And so in 1850 James Clerk Maxwell entered Peterhouse College in Cambridge. At one time this college had furnished Scotland with most of its chairs of Mathematics and Natural philosophy in her four Universities. But partly because of the smallness of the College he transferred after one term, which appeared to offer also a better chance of a Fellowship in the years to come. In the following year he was able to join the team of Hopkins, the great private tutor, as a fifteenth pupil.

During his time at Cambridge James Clerk Maxwell developed his theological viewpoint as well as increasing his mathematical skills. The theological viewpoint can be followed partly through his poem "A Student's Evening Hymn", which follows, and partly through his letters to Lewis Campbell in Oxford, which cover a range of dates.

A Student's Evening Hymn (1853)

Now no more the slanting rays
With the mountain summits dally,
Now no more in crimson blaze
Evening's fleecy cloudlets rally,
Soon shall Night from off the valley
Sweep that bright yet earthly haze,
And the stars most musically
Move in endless rounds of praise

While the world is growing dim,
And the Sun is slow descending
Past the far horizon's rim,
Earth's low sky to heaven extending,
Let my feeble earth-notes, blending
With the songs of cherubim,
Through the same expanse ascending
Thus renew my evening hymn
Thou that fill't our waiting eyes
With the food of contemplation,
Setting in thy darkened skies
Signs of infinite creation,
Grant to nightly meditation
What the toilsome day denies—
Teach me in this earthly station
Heavenly truth to realise.

* * *

Teach me so Thy works to read
That my faith—new strength accruing—
May from world to world proceed,
Wisdom's fruitful search pursuing;
Till, Thy truth my mind imbuing,
I proclaim the Eternal Creed,
Oft the glorious theme renewing
God our Lord is God indeed.

Give me love a right to trace
Thine to everything created,
Preaching to a ransomed race
By Thy mercy renovated,
Till with all thy fulness sated
I behold thee face to face
And with Ardour unabated
Sing the glories of Thy grace.

In a letter to Lewis Campbell written in 1851 he says "I believe with the Westminster Divines that 'Man's chief end is to glorify God, and to enjoy Him for ever', that to this end for every man He has given a progressively increasing power of communication with other creatures. That with his powers his susceptibilities increase. That happiness is indissolubly connected with the full exercise of these powers in their intended direction. That happiness and misery must inevitably increase with increasing power and knowledge. That the translation from the one course to the other is essentially miraculous, while the progress is natural."

As a necessary part of his skill in composition he read much. *Religio Medici* by Sir Thomas Browne was one of his favourite books. This was particularly true when he was laid up with a brain fever during the long vacation in1853. At this time his religious views were greatly deepened and strengthened. During this illness he was staying at the Rectory of Otley where he was invited by the uncle of a college companion. This friend has written, "Maxwell has left a very bright memory and example. We, his contemporaries at college, have seen in him high powers of mind and great capacity and original views, conjoined with deep humility before his God, reverent submission to His will, and hearty belief in the love and in the atonement of that Divine Saviour, who was his Portion and Comforter in trouble and sickness, and his exceeding great reward." Maxwell was profoundly moved by the kindness shown to him during this illness. He referred to it afterwards as having given him a new perception of the love of God. He knew then "that love abideth, though knowledge vanish away".

Graduation in Cambridge and Life Thereafter

In 1854 Maxwell graduated as Second Wrangler and equal Smith's Prizeman. These results were similar to those previously achieved by Wm. Thomson and Charles Mackenzie. He then began working at Electricity again, examining the heavy writings of his German contemporaries, Weber and Helmholtz. His comments on them are: "It takes a long time to reduce to order all the notions one gets from these men, but I hope to see my way through the subject, and arrive at something intelligible in the way of a theory." He also read Michael Faraday's *Experimental Researches*, as a result of which he was able to publish his first paper on the subject, "On Faraday's Lines of Force", which appeared in 1855.

In the following year he was appointed to the Chair of Natural Philosophy at Aberdeen University. This had the advantage that Aberdeen was nearer to Glenlair, where his father was ageing, than Cambridge was. Within two years he had married Katherine Mary Dewar, the daughter of the Principal of the University. It appears that Maxwell was pleased with the shift to Aberdeen. In a letter to a colleague he wrote, "My lines are so pleasant to me that I think everybody ought to come to me to catch the infection of happiness. This college work is what my father and I looked forward to for long, and I find we were both quite right, that it was the thing for me to do. And with respect to this particular college, I think we have more discipline and more liberty, and therefore more power of useful work, than anywhere else."

Thereafter he interrupted his electrical studies to explain the motion and permanence of Saturn's rings, for which he was awarded the Adams prize by Cambridge University in1859. In the following year the Chair at Marischal College was suppressed, but Maxwell was quickly appointed to the Chair of Physics and Astronomy at King's College, London. He then contracted smallpox at a fair near Glenlair, and was nursed by his wife in an isolated section of the house. His recovery from this disease was always later attributed by Maxwell to the care of his wife.

In 1861/2 Maxwell published his second substantial paper on Electricity, entitled "On Physical Lines of Force," and then a further paper in 1864 called, "The Dynamical Theory of the Electromagnetic Field." This unifies the separate electric and magnetic fields into a combined single electromagnetic field, in which the time changing electric field produces a time changing magnetic field, and conversely, ad infinitum.

Maxwell's love of Glenlair is believed to have been the cause of his resignation from King's College, London, in 1865. He wished to put into effect his father's plans for enlarging the house there. While this was being done the Maxwells took a trip to Italy, where James apparently became fluent in Italian. On their return James returned to his broad coverage of research, covering mathematics, thermodynamics, molecular theory and electricity and magnetism. Included in these were his *Treatise on Heat* and his *Treatise on Electricity and Magnetism* published in 1873.

Then in 1871 Maxwell returned to Cambridge as Director of the newly established Cavendish Laboratory. His work there was an outstanding success. But it did not include an attempt to produce electromagnetic waves. This was left to Heinrich Hertz, whose work will next be considered.

Heinrich Rudolf Hertz (1857-1894)

Heinrich Hertz was born in Hamburg of a wealthy family. His father was a barrister who later became a Senator. His mother had high expectations of her son, and expected him to be always first in his class at elementary school, which he entered at the age of six. These expectations he fulfilled. In addition to his formal schooling he attended lessons on geometrical drawing on Sundays. In such a cultured family it is surprising that he was totally unmusical, but he was provided with a workbench and woodworking tools at age twelve. Later he was to acquire a lathe also. At age fifteen he entered the Johanneum Gymnasium and came first in Greek, and while there he took private lessons in Arabic. He then went to Frankfort and worked for a year with an engineering contractor's firm. In the following year he attended Dresden Polytechnic for a time, before he had to do a year's military service, which he undertook with the railway regiment at Berlin.

By now in 1877 he was aged twenty, and went to study at the Technical High School in Munich. But he also matriculated at the University there, and followed this up by writing a most intriguing letter to his parents on 1/11/1877. He had decided that engineering was not the career he should follow, and the letter read as follows: "And so I ask you, dear father, for your decision rather than for your advice; for it isn't advice that I need, and there is scarcely time for it now. If you will allow me to study natural science I shall take it as a great kindness on your part, and whatever diligence and love can do in the matter, that they shall do." The father's decision was in accordance with his own wishes, and therefore in his first semester at Munich he studied Mathematics, reading the original papers of Lagrange, Laplace and Poisson. In his second Semester he balanced this theoretical study by gaining experience in laboratory work, both at Munich University and at the Technical High School. This work he found very satisfying.

In 1878 at age twenty-one he moved to Berlin to study under Helmholtz and Kirchhoff. This would seem to indicate that he had a clear objective for his future scientific life. Yet in November he wrote to his parents: "When I am only studying books I am never free of the feeling that I am a perfectly useless member of society." This was his experimental side coming out. In Berlin he saw that a prize was on offer from the Philosophical Faculty for the solution of a problem on "Electrical Inertia". This related to the extra current produced in a secondary circuit when a primary electric current starts or stops. Experiments on the size of this extra current had to be made so that a conclusion could be drawn as to the inertia in motion. This problem had been set by Helmholtz, and Hertz believed that it was not outside his line of interest, nor outside his ability to solve. Accordingly he discussed his proposal with Helmholtz himself, about how he would begin and what instruments he would require. Helmholtz then gave him a room to himself, to which he could come and go as he liked, and came in every day for a few minutes to see how things were progressing. Although the problem had been given out in the previous August, with the solution to be given by the following May, Hertz finished the project early in the New Year and obtained the Prize, a gold medal, in August 1879. When given the gold medal he remarked on "the incredible stupidity of the absence of any inscription, nothing even to show it is a University Prize." In addition to the prize of the gold medal, Hertz produced in 1880 his first research paper, of thirty-four pages, based on the results he had obtained.

Helmholtz then asked Hertz to try for the Berlin Academy Prize on "An experimental decision on the critical assumption of Maxwell's theory". This was a problem he had designed for his most talented student. But Hertz declined, considering that it would take him three years, and the outcome was uncertain. Instead Hertz wrote his doctoral dissertation. For this he turned to the subject of Moving Conductors, first in *Induction in Rotating Spheres*, which when published in 1880 occupied ninety-four pages, and was highly mathematical. The problem was completely solved for the case when the sphere is solid or hollow, rotating about a diameter. The inducing magnet may be outside, or in the case of a hollow sphere in the inside space. In the second associated paper, he dealt with "The distribution of Electricity over the Surface of Moving Conductors". In this he showed that the potential on the surface and also inside a rotating spherical conductor is no longer constant, and produced experimental confirmation of this. Thus a hollow conductor does not entirely screen its interior from external influences when it is in motion.

This work on Moving Conductors was submitted for a Ph.D in 1880, and the dissertation was awarded a "Magna cum Laude" acceptance, which was a rare distinction in Berlin. Also in 1880 Hertz became a salaried Assistant to Helmholtz for three years. This became the most productive research period in Hertz's life. In it he produced fifteen research papers. Their subject matter included the following with electrical titles:

(1) Hot wire ammeter of Small Resistance and Negligible Inductance.
(2) On a Phenomenon which accompanies the Electric discharge.
(3) Experiments on the Cathode Discharge.
(4) On the Behaviour of Benzene with respect to Insulation and Residual Charge.
(5) On the Relations between Maxwell's Fundamental Equations and the Fundamental Equations of the Opposing Electromagnetics.
(6) On the Relations between Light and Electricity.
(7) On the Passage of Cathode Rays through Thin Metallic Layers.
(8) On the Dimensions of Magnetic Poles in different systems of Units.

Number (5) above was a comparison between Maxwell's system and the opposing system which was based on direct action at a distance. He was able to show that "this system must introduce different kinds of electric force which it has never done, or it must admit the existence of actions which hitherto it has not taken into account." In this analysis Hertz introduces the concept of magnetic currents, showing the attraction between them, and their leading by a different route to the same results obtained by Maxwell. This paper was published in 1884, but did not meet with complete acceptance, possibly due to the concept of magnetic currents. It was only in 1888, and largely due to later work by Hertz, and the enthusiastic support of Fitzgerald, Helmholtz and Lord Kelvin that Maxwell's results became universally accepted.

After his surge of research papers in Berlin, together with the fact that his Assistantship to Helmholtz had terminated, Hertz looked for a regular faculty appointment. To achieve this he had to start as a Research Fellow, which was unsalaried. There were, apparently, too many of these in Berlin at the time, so Hertz went to the University of Kiel in 1883, recommended by Professor Kirchhoff in Berlin. There he had more free time, but he became depressed at age twenty-six. In a letter to his parents he referred to the sameness of his evening walk, though later in the year when he was on military exercises he wrote again to say he felt completely well at present. One of his problems was that Kiel University had no Physics Laboratory, so that he set one up in his own house. In his two years there he produced three research papers, one on meteorology, one on electric and magnetic units and the third on Maxwell's Electrodynamics. In 1885 the University offered him an Associate Professorship, which was salaried, but he declined this and went instead as Professor at Karlsruhe Technical High School. During the time when he was viewing the possibility of the two posts at Kiel and Karlsruhe he wrote to his parents, "I view the future through glasses so dark that I am subjectively convinced that neither a provincial Professorship at Kiel, or the possibility of a post at Karlsruhe will eventuate, and then torture myself with these spectres and monsters of my fancy." Clearly he was again having psychological problems.

But in 1886 he married Elisabeth Dull, the daughter of a colleague at Karlsruhe, and in the same year began the series of experiments that made him world famous. In the Karlsruhe Cabinet he came across the induction coils that were to enable him to produce both a high frequency transmitter and a resonant receiver. Both were necessary in the experiments he was to carry out to tackle Helmholtz's problem for the Berlin Academy Prize of 1879. And the experiments for both transmitter and receiver were carried out almost entirely experimentally.

The two spiral conductors had been used to illustrate mutual inductance between them with their planes parallel to each other. When the current in one spiral was started or interrupted a spark would be obtained across the terminals of the second spiral. But the production of sparks is commonly accompanied by variability. Theoretical considerations of sparks are too difficult for analysis, so Hertz turned to an experimental investigation of his spiral circuits. He unwound one spiral, breaking it in the centre and attaching the terminals of the coil there, so that the spark gap was in the middle. The wire was then straightened so that instead of a spiral one had a straight wire with the spark gap in the middle. This arrangement also produced sparks in the second spiral.

Further experiments were made on measuring the potential difference between two points on the straight wire. This involved placing a spark micrometre across the two points. But these new sparks could not be stopped even when a short circuit was placed across the new spark gap, or still more puzzling, when one of the wires in the spark measurement circuit was disconnected. By dint of repeated experiments Hertz ended up with a linear dipole transmitter, three metres long, with a separate rectangular loop 120cm by 80cm, placed parallel to the transmitting dipole and at a distance of up to 50cm from it. Experimentally Hertz found a resonance phenomenon in the separate rectangular loop which was excited by the discharge through the transmitting dipole. The tuned resonator thus became a way of indicating oscillations in the original linear discharge circuit.

Hertz did not initially suspect that the discharge circuit oscillated. But the frequency obtained was more than 100 MHz, about a hundred times higher than any oscillator previously used. Hertz "then felt that he had obtained entry to the Berlin-centred physics community". The actual frequency obtained depended on the length of the transmitting dipole, and to a small extent the size of the metal knobs which terminated this dipole.

The other problem that Hertz had to overcome was how to get the electric and magnetic fields to detach themselves from the wire and go free as Maxwell's waves. This he achieved by the use of the linear dipole in which the electric charges are spread out along the wire, rather than being close together as in a capacitor. The reception of this wave was accomplished by an identical dipole with another spark gap at its centre, and with its orientation parallel to that of the transmitting dipole. The presence of waves was confirmed by means of a reflecting sheet, which reinforced the wave amplitude at points where the direct and reflected waves were in phase, and cancelled when they were in antiphase. In addition the waves could be bent out of their course by using a prism made of pitch, and would pass through walls and floors. Hence Maxwell's theory was vindicated, and Helmholtz's mind was clarified.

The experiments had occupied Hertz from 1886-1888. After these were completed he turned to explaining his results theoretically, and these results were published in two papers, in 1890, in his book *Electric Waves*. His own professional position was improved when he succeeded R. J. E. Clausius in 1889 as Ordinary Professor at the University of Bonn. In this position his research interests dealt with "Electrical Discharges through Gases"—but his health deteriorated through a malignancy in his mouth, and he died on 1st January 1894.

Hertz's father was a Jew who converted to Christianity, and Hertz himself was brought up as a Lutheran. Nevertheless his burial took place in the Jewish cemetery just outside Hamburg. In the 1930's when Nazi persecution of the Jews took place it is recorded that Hertz's wife and her two daughters felt it necessary to leave Germany. They then settled in Cambridge, England.

It was an obituary notice of Hertz's life which came to the notice of Guglielmo Marconi, and which captivated his interest, that set him along the path of developing both long and short wave radio propagation.

CHAPTER EIGHT

LODGE : TRUTH AND A REPUTATION TO LOSE

c.1850 – c.1900 (Section 2)

IN THIS CHAPTER we shall deal with the contributions made by the followers of Maxwell in transforming the writings of his theoretical two-volume treatise into the practical performance of producing radio waves. Although, as seen in Chapter Seven, this was accomplished by Hertz in 1889, the initial key suggestion was put forward six years earlier by G. F. Fitzgerald of Dublin. His proposal was to discharge a Leyden Jar through a small resistance.

After this the major engineering contributions were made by Guglielmo Marconi, the son of an Italian father and an Irish mother, who lived in a large villa near Bologna. It was in the attic of this villa that Marconi carried out his first experiments on radio. Although Marconi was unsuccessful on academic grounds in gaining entrance to the University of Bologna, his mother was able to persuade Professor Righi to lend her son some equipment from time to time, which enabled Guigelmo to pursue his experimental work. She used to bring a tray of food up to the attic when he was working long hours, and leave it outside the locked door of his laboratory. The achievements of Marconi were based initially on an accidental experimental observation made by himself.

With his radio link set up outside the villa in 1895, he records that *by chance* he held one of the plates of his transmitting antenna at a considerable height, and set the other on the ground. With this arrangement the signals became so strong that they gave a range of one kilometre, an improvement of one hundred times. Then the upper metal plate was raised still higher, with the lower one still on the ground. The plates were replaced by several copper wires at both ends, and still greater distances were achieved. At the end of

1895 Marconi discovered that his radio waves could go over a hill, and still give a range of approximately one mile. Thus one of the advantages of radio waves over light waves was found experimentally.

In September 1901 a 200-foot diameter ring of masts, 200-foot high was built at Poldhu in Cornwall, but in the same month it collapsed during a storm. Accordingly a simpler aerial was used as a replacement, where the wires took the shape of a fan. On the other side of the Atlantic, the Cape Cod aerial also collapsed during a storm in November 1901. But by using kites for the receiving aerial, transmission was successful from Poldhu to Cape Cod on 12th December 1901.

Prior to Marconi's transatlantic success in radio communication, Fitzgerald wrote to Oliver Heaviside in 1899 to ask, "Have you worked at the propagation of radio waves round a sphere? A case of this is troubling speculators as to the possibility of telegraphy by electromagnetic waves to America. It is," he said, "evidently a case of diffraction, and I think must be soluble." But Heaviside did not take the problem up, and in later years referred to this type of work as "those awful diffraction calculations".

Heaviside, who was a nephew of Charles Wheatstone, had started work on the experimental side of line telegraphy at age eighteen, in 1868, in Denmark, working on the first Anglo-Danish Telegraph Line which had been laid that year. He and his brother Arthur had the ambition to build a high speed telegraph system which could transmit electrical signals perfectly. For this they realised a mathematical theory was required, and in 1873, the year of its first publication, Heaviside acquired a copy of Maxwell's *Treatise on Electricity and Magnetism*. Later he said of the book, "It was great, greater and greatest." The following year Heaviside, who suffered from deafness, moved to London with his parents. While there he began his mathematical work each night at 10 p.m. when the house was quiet, and did his experiments during the day.

Some explanation appears to be required here, because in Denmark he had been employed by a commercial organisation, and now, in London, he had no income. The full explanation appears to be difficult to come by. One explanation is that he had offended the Chief Engineer of the British Post Office by belittling the existing theory, which he believed had been improved by himself. The other possibility is that he was showing early signs of being the recluse which he later became.

The fourth member of the group to be considered in this chapter is Oliver Lodge. He came from the potteries where his father had his own business selling clay to industrial concerns. For several years after leaving school he worked with his father, as a result of which he developed the personal skills which were later to be a bonus in the wider responsibilities he was to develop.

In 1872 the Government offered scholarships for young teachers at the Royal College of Science in South Kensington. Oliver Lodge was offered the last place on the course. This was in Biology under Huxley, but after a week or so it was possible to arrange a transfer to Chemistry. But Oliver found it possible to attend lectures in both subjects, in addition to fulfilling all his laboratory requirements. Beyond all this he attended further classes at King's College, where he took physics, mathematics, zoology and geology. Chemical analysis was the main feature of the practical course in South Kensington, and Oliver soon attracted the attention of the teaching staff because of the accuracy of his results. As for accommodation he was able to stay with his mother's sister Anne, as a follow on to an earlier stay with her, which Anne had arranged to include preparation for his Confirmation.

During his Scholarship Year in London, Oliver had the good fortune to be able to attend Lectures given by the Royal Institution Lecturer Professor Tyndall, on the subject of heat. Oliver found the lectures were brilliantly delivered and well-illustrated. As a result he attended as many as possible of other lectures given at the Royal Institution in their Friday Evening series.

Attention will now be turned to providing more details of each of the four major contributors, Fitzgerald, Marconi, Heaviside and Lodge on the development of radio communication between c.1875 and c.1900.

George Francis Fitzgerald (1851-1901)

His father had held a Chair in Trinity College, Dublin, before becoming a leading Churchman in the Church of Ireland, first as Bishop of Cork and later of Killaloe. His mother was the sister of George Christopher Stoney who invented the word "electron" commonly used. As a boy he did not go to school, but was tutored by the sister of George Boole (1815-1864). George Boole's father was an impecunious shoemaker who lived in Lincoln, but George, through self-study, became famous for his book *The Mathematical*

Analysis of Logic. As a result he was appointed as the first Professor of Mathematics at University College Cork in 1849.

George Francis Fitzgerald entered Trinity College, Dublin and graduated from there in 1871 with the top place in Mathematics and Experimental Science. Then it was a case of waiting for a vacant Fellowship in Trinity. This took him six years. During this time, after the manner of the pick of the Dublin men, he settled down to a wide and independent course of reading for the Fellowship. The examination in mathematical and physical science included papers on selected portions of the works of the great mathematical physicists. As was said in an obituary in *Nature* of Fitzgerald's life, "Acquaintance with the present state of science, however detailed and exact, assumes its full value as an instrument of progress only when it is accompanied by appreciation of the difficulties that had to be circumvented in order to reach it, and by observation of the way in which complete logical precision may have to be attained at the expense of temporary limitation."

In the year after his appointment as Fellow of Trinity College, the British Association held its annual meeting in Dublin. Fitzgerald was twenty-seven years of age, and it was the first meeting he had attended of this Association. It was here that he met for the first time Oliver Lodge, who was of the same age, and who had been a devoted follower of the B.A. for many years. The two men hit it off well, both being very able, but in different ways. Fitzgerald was an analyst, while Lodge brooded over problems, looking for a mechanical model as an analogue to the cause of his perplexity.

Their friendship lasted all Fitzgerald's life, and involved visits to their respective departments in Dublin and Liverpool where Lodge had been appointed to the first Chair of Physics there in 1881. Similarly Fitzgerald was appointed to the Erasmus Smith Chair in Dublin in the same year. In their correspondence with each other they adopted the practice of signing their names with the capital Greek letters Phi for Fitzgerald, and Lambda for Lodge, in a similar way to Thomson and Tait signing themselves T and T'.

In his new post Fitzgerald began the teaching of practical physics, and campaigned for it throughout the whole of Ireland. One of his demonstrations required an attempt to fly the Lilienthal Glider. This was unsuccessful. The verbal student response was to refer to him thereafter as Flightless Fitzgerald, but this was unfair. Lilienthal himself was killed in a later attempt at flying his own design of glider.

In Mathematical Physics it became the practice to teach from the great writers, Newton, Lagrange, Laplace, Poisson, Gauss, MacCulloch, Ampere, etc. instead of using the recastings of textbook writers. This practice appears to offer many advantages for the more able student, especially for those who later proceed to research. Perhaps associated with this is the fact that Fitzgerald himself never wrote a book. His best ability was his outpouring of ideas on many subjects. And this was associated with an absence of any desire to claim priority for such conceptions.

With such a gifted Fellow of Trinity College it was perhaps inevitable that there might be tensions within the Senior Common Room. Two contradictory stories may therefore combine to give the best report possible on the atmosphere within it. The first relates to a visit by Oliver Lodge in 1878 when he saw the great men of Trinity having breakfast with the Provost Dr Salmon. He records that the Senior Fellows seemed to spend their time in telling comic tales, each one capping the other, and not having much serious conversation. In contrast to this the folk tales of Trinity include one in which Fitzgerald pulled a knife on another Fellow whose inanities he could no longer stand.

In his short life of forty-nine years Fitzgerald produced over one hundred research papers. But it is not the number, so much as the breadth of interests he covered, which astonishes the research worker of today. The following list of about one quarter of these titles is given as an illustration:

(1) "On the effects of Magnetisation of the Iron in a Ship, on the Compass, When the ship Heels", *Hermathena*, 1876.

(2) "On the Rotation of the Plane of Polarisation of Light by Reflection from the Pole of a Magnet", Proc. of the Royal Society, 1876.

(3) "On the Electromagnetic Theory of the Reflection and Refraction of Light", Phil. Trans. of the Royal Society, 1880.

(4) "Note on Surface Tension", British Association Report, 1878.

(5) "On the Possibility of Originating Wave Disturbances in the Ether by Means of Electric Forces", Parts 1 and 2, Royal Dublin Society, 1879 and 1880.

(6) "Corrections and Additions to Above", Royal Dublin Society, 1882.

(7) "On Comets Tails", Royal Dublin Society, 1882.

(8) "On Electromagnetic Effects due to the Motion of the Earth", Royal Dublin Society, 1882.

(9) "On the Quantity of Energy Transferred to the Ether by a Variable Current", Royal Dublin Society, 1883.

(10) "On Maxwell's Equations for the Electromagnetic Action on Moving Electricity", British Association Report, 1883.

(11) "On the Energy lost by Radiation for Alternating Electric Currents", British Association Report, 1883.

(12) "On a Method of Producing Electromagnetic Disturbances of Comparatively Short Wavelengths", British Association Report, 1883.

(13) "On a Method of Studying Transient Currents by means of an Electrodynamometer", Royal Dublin Society, 1884.

(14) "On a Non-Sparking Dynamo", Royal Dublin Society, 1884.

(15) "Sir William Thomson and Maxwell's Electromagnetic Theory of Light", *Nature*, May 7, 1885.

(16) "On the Limits to the Velocity of Motion of the Working Parts of Engines", Royal Dublin Society, 1886.

(17) "Experimental Science in Schools and Universities", *Nature*, Vol. 35, 1886.

(18) "On the Accuracy of Ohm's Law in Electrolysis" (Joint with F. T. Trouton F.R.S.), British Association Report, 1888.

(19) "Address to the Mathematical and Physical Section of the British Association in Bath", British Association Report, 1888.

(20) "Multiple Resonance Obtained with Hertz's Vibrators" (Joint with F. T. Trouton F.R.S.), *Nature*, January 30, 1890.

(21) "Electromagnetic Radiation", Proc. of the Royal Institution of Great Britain, 1890.

(22) "An Estimate of the Rate of Propagation of Magnetisation in Iron", *Nature*, August 10, 1892.

(23) "Magnetic Action on Moving Electric Charges", *Electrician*, July 14, 1899.

(24) "Sunspots, Magnetic Storms, Comets' Tails, Atmospheric Electricity, and Aurorae", *Electrician*, December 14, 1900.

(25) "Mr Hargrave's Paper on Sailing Birds", *Aeronautical Journal*, July 1900.

But it should not be thought that Fitzgerald always got things right at his first attempt. His 1880 British Association paper at Swansea can be summarised by a one-sentence line: "Fitzgerald by comparing Maxwell's Theory to Action at a Distance Theory had deduced the conclusion that electric currents and systems cannot originate in the ether such distributions as those of light." William Ayrton found his argument unconvincing. His partner John Perry thought that electromagnetic production of light seemed a probable occurrence. Then Fitzgerald found a flaw in his own proofs by studying Lord Rayleigh's *Theory of Sound*. Rayleigh had dealt with the same differential equation he had found in 1880. But Rayleigh gave a different solution corresponding to a train of travelling rather than the standing waves Fitzgerald had found. In 1882 Fitzgerald in the Royal Dublin Society pointed out that both solutions were valid, but his own implied the presence of a reflecting sheet. Rayleigh had also used retarded potentials to treat the propagation of sound waves. Fitzgerald then used a similar retarded potential to calculate the energy in the radiated wave.

George Francis Fitzgerald as a Man: The Views of his Contemporaries

Because Fitzgerald was a doyen of Natural Philosophy, it seems appropriate to begin this section with the views of another leader in the same field, namely Oliver Heaviside. After Fitzgerald's death Heaviside wrote as follows:

> "I only saw him twice knowingly, once for two hours and then again for six hours, after a long interval; yet we had a good deal of correspondence at one time, and I seemed to have quite an affection for him. A mutual understanding had something to do with that. You know that, in the pre-Hertzian days, he had done a good deal of work, not large in bulk, but very choice and original in relation to the possibilities of Maxwell's theory, then considerably undeveloped and little understood; and his way of looking at things was more like my own than anybody's. Well, he found that I had done a lot of work in the same line, and he was most generous in recognising and emphasising it. Too generous, of course. You remember that view of my 'Electrical

Papers' that he wrote? No one knew better than myself how to allow for his temperament and desire to help me. He used to write to me a good deal about electromagnetic problems, and I laid down the law to him like—like myself, in fact. He took it all very pleasantly. But I knew all the while that he had a wider field than myself, and no time to specialise much. He had, undoubtedly, the quickest and most original brain of anybody. That was a great distinction; but it was, I think, a misfortune as regards his scientific fame. He saw too many openings. His brain was too fertile and inventive. I think it would have been better for him if he had been a little stupid—I mean not so quick and versatile, but more plodding. He would have been better appreciated, save by a few."

From the scientific journal *Nature* comes the following tribute:

"The scientific public of this country was placed very early in touch with Hertz's magnificent and decisive verification of electrodynamic theory, through the attention commanded by Fitzgerald's brilliant exposition in his British Association address of 1888. It was fitting that this should have come from him; for as Lord Kelvin has recalled, he had five years before [he] pointed out to the British Association the possibilities of the very plan of obtaining electric radiation of manageable wavelength, which in Hertz's hands has led to success.—He followed very closely the progress of abstract mathematical physics; hardly anyone could be named who had thought more deeply, or whose knowledge was more available and many-sided, more entirely free from all prepossession or prejudice. His stores of knowledge were ripening and maturing year by year; his memory was unfailing, and each new fact or phenomenon seemed to find its place at once in the setting to which it belonged."

Because of his breadth of interests it is not surprising that there were tributes from scientists in other fields. Thus Professor W. Ramsay, a co-discoverer of argon, neon, krypton and xenon, wrote:

"The blow is so recent and the feeling of personal loss is so acute that this is a difficult task. But to me as to many others, Fitzgerald was the truest of true friends; always interested, always sympathetic, always encouraging, whether the matter discussed was a personal one, or one connected with science or with education. And yet I doubt if it were these qualities alone which made his presence so attractive and so inspiring. I think it was the feeling that one was able to converse on equal terms with a man who was so much above the level of oneself, not merely in intellectual qualities of mind, but in every respect. I know that Fitzgerald himself would have been the last to acknowledge this, for he had no trace of intellectual pride; he never put himself forward, and had no desire for fame; he was content to do his duty. Generally he saw the best in people, and like Lord Kelvin was able to disentangle ideas of value from the crude efforts of presentation of a beginner or of an ordinary muddle-headed man."

As Sir Oliver Lodge put it:

"The idealistic turn of his mind in dealing with ultimate questions came out constantly in his conversation on such topics, and may be illustrated by a quotation from the end of his Helmholtz Lecture. After noting that all forms of external stimulus, into whatever terms we translate them—sound, colour and the rest, nay even space time and substance too, perhaps—resolve themselves into motion, he goes on to ask: And what is the inner aspect of motion? In the only place where we can hope to answer this question, in our brains, thought turns out to be the internal aspect of motion. Is it not reasonable to hold, with the great and good Bishop Berkeley that thought underlies all motion?—For the highest life we require the highest ideal of the Universe to work in. Can any higher exist than that, as language is a motion expressing to others our thoughts, so Nature is a language expressing thoughts, if we learn but to read them."

On a lower level than this we know that after J. J. Thomson had been appointed to the Cavendish Chair in Cambridge, Fitzgerald wrote to him congratulating him on his appointment, for which he himself had applied. He wrote: "I hope you will succeed in getting your experimental test of Maxwell's theory tried. The great difficulty is to find those rapidly alternating currents. Would Langley's bolometer do?" (This was a thermocouple device). His own calculations had suggested the effect would not be measureable. Even today virtually all detectors use a resonance effect to enlarge the week signal. But neither Fitzgerald nor Lodge built such a detector before 1888.

One of his colleagues on the literary side said:

> "His appearance was not unworthy of his fame. More striking he was than handsome, but his ample grey locks and beard, his furrowed brow, his penetrating eyes, reminded one of the bust of some Greek philosopher, which we cannot look upon without that feeling of respect which intellect and character command among civilised men."

We now turn to the next theoretician, Oliver Heaviside, an Englishman born in London in May 1850.

Oliver Heaviside (13/5/1850—3/2/1925)

His father was a skilled wood engraver who had come to work in London from Stockton-on-Tees. The trade was going through a difficult time because of the introduction of photographic etching techniques and Oliver's mother found it necessary to start a school for girls in North London. Oliver lived in this neighbourhood for thirteen years, and hated it. But the sight of the drinking in the pub, which was opposite his home, made him a teetotaller for life. His social home environment also left much to be desired, possibly because of his father's anxiety over the absence of work, and the associated stress on his mother, coupled with her own worry in running a school. But Oliver was a great reader, particularly of Dickens's novels, a pastime not shared by other boys in the neighbourhood, with whom he had no contact because of his own deafness, brought on by an attack of scarlet fever.

When Oliver was thirteen, in 1863, the Heaviside family moved to a better part of North London. He went to the local grammar school and took the College of Preceptors examination there at age fifteen. The thirteen subjects he sat included English, Latin, French, Physics, Chemistry and Mathematics, and he won First Prize in Natural Science. The following year he left school, but it is not clear what he did for the next two years. But he began working as a Telegraph Operator, in 1868, on the first Anglo-Danish Telegraph Line which had been laid in September 1868. In this connection it is relevant to say that his mother's brother was Charles Wheatstone (1802-1875), who started life as a maker of musical instruments, and later published a paper entitled "New Experiments in Sound." He became Professor of Experimental Philosophy in King's College, London, in 1834, and took out the first patent, in conjunction with W. E. Cooke, for the electric telegraph in 1837. The relationship between Wheatstone and the Heavisides must have been good, because of the four Heaviside brothers; one became the owner of a music shop and the other three worked in telegraphy.

After two years' experience in Denmark Oliver transferred to Newcastle on Tyne in 1870, where he was made Chief Operator, and was responsible for fault localisation on the submarine cable. This work involved familiarity with Kirchhoff's Circuit laws. Oliver and his brother Arthur, who also worked in Newcastle, had the ambition to build a high speed telegraph system which could transmit electrical signals perfectly. For this they realised a mathematical theory was required, and in 1873, the year of its first publication, Oliver acquired Maxwell's *Treatise on Electricity and Magnetism*. Later he said of this book, "It was great, greater and greatest."

The following year he moved to London with his parents. The reasons for this may be complex. In the previous year he had written a paper on a means of being able to transmit and receive a signal simultaneously. This is referred to as duplex operation of the system. In this paper he had criticised existing theories which were used in the British Post Office, led by the Chief Engineer, W. H. Preece. Preece was understandably annoyed, and this may have been responsible for a request by Oliver for a salary increase being rejected. However, it may also be true that Oliver may have been showing early signs of an isolationist tendency which developed strongly later.

The result was that Oliver stayed in London from 1874 to 1889. It will be recalled that he was previously in employment, so that his financial circumstances were now materially altered. But during this time he continued to educate himself, and in the later years he was able to publish papers in the *Electrician* for which he was paid £40 per annum from 1882 to 1887. The London years were scientifically the most productive in the whole of his life. He later said that he had done all his original work by 1887, and it was all contained in his two-volume *Electrical Papers* which were published in 1892. During the whole of these years his sole scientific collaborator was his brother Arthur, who was an experimentalist.

Heaviside, like Newton, Laplace and Ramanjan, had the facility of being able to discover mathematical theorems of a high order of difficulty and complexity, by some sort of intuition which dispensed with the usual process of proof. This meant that his work was extremely difficult to follow. Even a renowned physicist like Fitzgerald in reviewing Heaviside's *Electrical Papers* said, "In his most deliberate attempts at being elementary, he jumps deep double fences, and introduces short-cut expressions that are woeful stumbling blocks to the slow paced mind of the average man, when arguing about concepts that the average man is not very familiar with."

Likewise Hertz, in private correspondence, says, "I find it so difficult to follow your symbols and your very original mode of expressing yourself. Mathematical symbols are like a language and your writing is like a very remote dialect of it." Nevertheless there was friendly correspondence between Hertz and Heaviside, and on another occasion Hertz suggested gently to him that he ought to make his papers much simpler, much easier to read. But this recalls the story of someone who said, "You know, Mr Heaviside, your papers are very difficult indeed to read," when Heaviside's famous reply was, "That may well be, but they were much more difficult to write."

This difficulty of Heaviside's in explaining things simply was well put in a Presidential Address to the Institution of Electrical Engineers when the President said, "Mr Heaviside was unfortunately not imbued by nature with the facility for an exposition in a diluted form." One had to dig deep to see what he was saying. But we have to remember that as Heaviside regarded himself as the apostle of Maxwell, he had to dig deep in reading Maxwell's *Treatise* to be able to find out how it could be applied to practical problems. It has been pointed out that he then set himself the task of evolving a new point of view, without impairing the accuracy of Maxwell, and he then concluded, "I then put Maxwell aside and followed my own way... And I advanced much faster."

Examples of Heaviside's Thinking

The Rational Current Element

It will be recalled that Ampere analysed the magnetic field set up by a closed loop of electric current. From this analytical result he deduced the strength and direction of a current element Idl. in a way that was not rigorous. What Heaviside did was to imagine this current element in the form of a short cylindrical torch battery, with the usual insulation round its curved sides, placed in an infinite ocean of seawater. Current flows out radially and uniformly from the positive pole and converges in the same way into its negative pole. The current flow is symmetrical about the axis of the battery and so the magnetic field takes the form of circles centred on this axis. He found the magnetic field at any point in the ocean in terms of the length of the battery and the total current flowing through it. Though current is flowing everywhere throughout the ocean the magnetic field is described in terms of the current flowing in that one place, where all of it is constrained to gather together and pass through the battery.

Bending of Waves

In the tenth edition of the *Encyclopaedia Britannica*, Heaviside wrote an article in 1902 on the transmission of radio waves along a pair of wires, showing that the waves can be guided round a curved path provided the path changes direction slowly. He says, "This guidance is obviously a most important property of wires. There is something similar in wireless telegraphy. Seawater, though transparent to light, has quite enough conductivity to make it behave as a conductor for Hertzian waves, and the same is true in a more imperfect manner of the earth. Hence the waves accommodate themselves to the surface of the sea in the same way as waves follow wires. The irregularities make confusion, no doubt, but the main waves are pulled round by the curvature of the earth and do not jump off.

"There is another consideration. There may possibly be a sufficiently conducting layer in the upper air. If so, the waves will, so to speak, catch on to it more or less. Then the guidance will be by the sea on one side and the upper layer on the other."

Artificial Loading of Transmission Lines

In 1891 John Stone, Stone of Bell Telephone Company, wrote to Heaviside for advice regarding the elimination of distortion in telephone lines. Heaviside analysed the problem, bringing in inductance L and conductance G, resistance R and capacitance C, all per unit length, and showed that when R/L equalled G/C the transmission is distortionless. However, after this promising beginning, the first U.S. patent for distortionless telephone lines, designed according to Heaviside's general theory, was awarded in 1901 to Professor Michael Pupin of Columbia University. The invention brought Pupin hundreds of thousands of dollars by the mid-1910's. Heaviside never recovered from the shock of being deprived of the rights of an invention he considered his own.

Heaviside's Personal Characteristics

Election to Fellowship of the Royal Society

In 1891 Sir Oliver Lodge asked Heaviside's permission to put his name on the candidate's list for election to the Royal Society. Heaviside replied he would only agree if given a guarantee of election on the first go. He said if he were considered and not elected he would make a public row over it. He was elected, but both Silvanus Thompson and Joseph Larmor, both candidates at the same time, had to wait for another year.

After the election he was asked to go to London to be admitted, but he would not do that. He wrote a short poem instead, complaining about the £3 fee asked for. The poem goes as follows:

> Yet one thing more before
> Thou perfect be.
> Pay us three pound,
> Come up to Town
> And then admitted be
> But if you won't
> Be Fellow, then Don't.

Award of Honorary Ph.D from University of Gottingen

The University of Gottingen is famous for having the largest number of Nobel Prize Winners among its former staff and student body than for any other city except for Stockholm. It is particularly pleasing that the University which numbered Carl Gauss among its staff should honour Oliver Heaviside in this way. No other University is known to have offered Heaviside an Honorary degree. The translation from the Latin of this Diploma has been freely rendered in the following way:

> That Eminent Man
> Oliver Heaviside
> An Englishman by Nation Dwelling at
> Newton Abbott
> Learned in the Artifice of Analysis
> Investigator of the Corpuscles
> Which are wont to be called Electrons
> Persevering, Futile, Haply Though
> Given to a Solitary Life
> Nevertheless among the Propagators of the
> Maxwellian Science
> Easily the First.

Nomination for Nobel Prize in Physics

In 1910 Heaviside was nominated for the Nobel Prize in Physics, but was not successful when the awards were announced. Instead it went to J.D. van der Waaals "for his work concerning the equation of state of gases and liquids".

First Recipient of Faraday Medal: Heaviside (1921)

On the occasion of this award the record of the Institution of Electrical Engineers is: "They are convinced that, as now, so in the future, the name of Heaviside will rank among those of the great founders of the science of Applied Electricity."

Heaviside in a Nutshell

To Heaviside belongs the credit for reducing Maxwell's eight electrical quantities (voltage, magnetic scalar potential, magnetic vector potential, electric field strength, electric current, electric displacement, magnetic flux density, magnetic field strength) down to two (electric field strength, magnetic field strength). Much of his analysis was based on transmission lines, and this work on lines can be translated to the field of radiowave propagation over the earth, simply by separating the plates of a parallel plate transmission line. When the separation distance is increased to infinity one is left with linear propagation over a lossy ground.

Heaviside was an eccentric, a man of genius who did not fit into society. It is said that he kept and filed all his correspondence, except for bills, which he often threw away without paying. Consequently his gas supply was once cut off for about one year. In this sense he resembles the poet Francis Thompson, who also knew privation, and once lived for a week on sixpence earned for holding a horse's head. When a kindly neighbour saw Heaviside looking very cold and ill in the garden, she told him to go inside and sit by his fire. "Madam," replied Heaviside with a smile, "I have no fire—I have only my genius to keep me warm."

Heaviside postulated the existence of magnetic monopoles and currents, although there was no experimental evidence that such entities did or could exist. One reported successful finding occurred on 14th February 1981 (St. Valentine's Day), but this was not universally accepted as a valid result. As an illustration of this scepticism when 14th February arrived in the following year, one of the sceptics sent a telegram to the supposed discoverer which read: "Roses are red, violets are blue, The Time is Now for Monopole Two."

After that brief summary of Heaviside, attention will now be turned to the third of the men in the U.K. who helped to make Maxwell's views become what they now are, universally accepted and used.

Oliver Lodge (1851-1940)

Both of Oliver Lodge's grandfathers had been headmasters. In addition one of them had been a vicar as well as being headmaster. Oliver's father worked for some years in the offices of the North Stafffordshire Railway in Stoke.

His mother had a sister who became Woman of the Bedchamber to the widow of King William IV. Oliver himself learned to read very early, partly through the help of his grandmother, who paid particular attention to the importance of reading aloud. As a result he became an excellent public speaker later in life.

At the age of eight he was sent to a boarding school at Newport in Shropshire, where a young clergyman, who was his uncle, had been given the position of second master, after transferring to a secular life. His comment on his schooldays from age eight to fourteen many years later, was that they were the dullest and most miserable that he could easily picture. This was despite his transfer in 1863 to Combs Rectory, to which his uncle had transferred, but who wished to supplement his income by taking a small number of private pupils.

At about the same time Oliver's father had set up in business on his own, supplying clay to Staffordshire potters. After this transfer Oliver had a holiday in Edinburgh with his father, who was recuperating there after a fall from a horse. In addition he had a holiday in Weston-super-Mare with his Aunt Anne, who developed his interest in astronomy by encouraging him to build an astronomical model. Then at age fourteen Oliver was withdrawn from school to learn the trade which his father hoped would soon be called Lodge and Son.

It was the influence of Oliver's Aunt Anne again that made his parents accept the fact that Oliver's spiritual and mental needs were being neglected. As a result it was decided that Oliver should spend the winter in London with his Aunt Anne, have lessons to prepare him for confirmation, and attend classes at University College and King's College. On Sundays they both attended church and both took notes on the sermons delivered.

With regard to the University classes Oliver was almost overwhelmed by the quality of Professor Tyndall's lectures on heat. As a result he did all he could to obtain tickets for any of his Royal Institution Lectures on Friday evenings. After his return home to the Potteries, Oliver enrolled for classes in physics and mathematics in particular, and carried out experiments with batteries and galvanometers in an outhouse. These studies occupied the years 1868-1872.

Oliver then became aware in 1871 of the London University External Degree. This required Matriculation, followed by Part I and Part II examinations. The subjects for Matriculation were all compulsory, Latin,

Greek, another foreign language, English, Geography, History, Mathematics, Natural Philosophy and Chemistry. Oliver worked on his own with only a set of past papers to guide him and passed in the First Class.

In the following year he sat for Part I but was unsuccessful. Nevertheless he continued his studies and achieved a first class pass in 1873. He then tried for an exhibition at St John's College, Cambridge, but came second. Because he was aware of the excellence of the teaching in Mathematics at University College, he went there in 1874, and ended up assisting George Carey Foster in Physics which provided him with a payment of £50 p.a. This work he continued for three years until he got married. Part II was then passed, and was followed by the London DSc in June 1877. Oliver then remained at University College until 1881 when he moved to Liverpool.

In order to gain recognition in Physics it was necessary to publish research papers, and Oliver's first important paper was written with Carey Foster in 1875. It dealt with the shape of the lines of current flow passing between two electrodes on a plane metal sheet. In fact this problem had previously been solved by Kirchhoff and Robertson Smith who had later been dismissed by Aberdeen University on a heresy charge. Later Robertson Smith went to Cambridge and among other work edited *The Encyclopaedia Britannica*.

Oliver Lodge first attended a meeting of the British Association in Bradford in 1873 when he was still working for his father. Over a period of more than fifty years he did not miss more than two or three meetings. His love of speaking in public, linked with the tradition that the papers should be intelligible to the non-specialist, no doubt helped to contribute to these discussions. One particular paper was given by Alexander Graham Bell, whose beautiful enunciation impressed Lodge tremendously.

About this time Oliver Lodge became engaged to Mary Marshall, the girl whom he had known all his life in the Potteries. His fiancée's parents were, like Oliver's father, unhappy about his financial prospects in a University. This resulted in an ultimatum that the marriage could not take place until Oliver provided evidence of being able to earn £400 per annum. Through teaching at Bedford College in addition to University College, and taking on much marking, he was able to satisfy more than the requirement that had been laid down.

At the age of twenty-nine he published his first book, *Elementary Mechanics, including Hydrostatics and Pneumatics*. This set out in simpler form the ideas contained in conventional textbooks.

In 1882 Oliver became acquainted with two people who were later to transform his life, Edmund Gurney and F. W. H. Myers, the founders of the Society for Psychical Research.

Their friendship, however, had no special significance while Oliver was working in London. But an opportunity for promotion from assistant professor to full professor arose when Owen's College in Manchester advertised a Chair in Applied Mathematics. This was to be in mathematical physics as a complement to the work of Balfour Stewart who was an experimentalist. The chair went to Schuster who was already teaching in the department. He was of independent means and was already paying the salary of a reader in mathematical physics.

At the same time in 1881 another appointment was advertised for a Professorship in Experimental Physics and Mathematics in the new University College at Liverpool. This was to be formed around the existing Medical School, so that initially the work would involve providing the physics and mathematics needed by the medical students. Oliver Lodge was offered this appointment without interview, on the strength of the excellent testimonials he had accumulated. The salary was to be £400 p.a. and permission was granted to tour the Universities of Europe with the object of making use of good ideas which might have originated, or been found to be useful there. In Berlin he met Helmholtz, but was shown around that department by Heinrich Hertz, who was soon to become famous for his work on electromagnetic waves, and to be a collaborator with British and Irish scientists.

In 1884 Lodge joined the Society for Psychical Research. There were about seven hundred members including sixty university dons and eight Fellows of the Royal Society. By 1924 some of the older members had acquired titles, Orders of Merit and Nobel Prizes. More will be said about this activity later.

One of Lodge's main research activities was concerned with electromagnetic waves. At that time there seemed to be no connection between traditional sources of light, such as flames, or the sun, or moon, or stars, on the one hand, and electricity. But Fitzgerald, who was Lodge's friend, speculated that an electric current could produce an electromagnetic wave, and in 1883 wrote a paper to that effect. And his proposal is still used in most radiocommunication antennas. But no effort was immediately made to prove this experimentally.

In 1888 Lodge gave lectures on lightning conductors in which he made use of the Leyden Jar to produce spark discharges to display the part of lightning flashes. The central theme of Lodge's theory was that the lightning flash consisted of an oscillatory current, and not a direct current as maintained by more practical engineers, such as Mr William Preece of the Post Office. The dispute between the two sides was fairly bitter, producing this remark from Oliver Heaviside:

> "Although Mr Preece, in the presence of some distinguished mathematicians recently boasted that he made mathematics his *slave*, yet it is not wholly improbable that he is a very striking and remarkable example of the opposite procedure."

Also in 1888, using the discharge of a condenser as Fitzgerald had suggested, Lodge succeeded in producing electromotive waves which he sent along wires and obtained reflections and interference effects similar to those found with light. But in 1887 Fitzgerald had been forestalled by Hertz in Germany. Hertz had used an induction coil to produce a spark, and at the receiver some distance away a minute spark was also produced.

In late 1889 Lodge had his first personal experience of psychical phenomena. Mrs Piper was an American who had been invited to come to this country to demonstrate such phenomena. In Cambridge Lodge had his first sitting with a trance medium. During the course of it there was an electrifying change in his attitude towards such experience. It appeared to Lodge that his Aunt Anne, who had died of cancer, took possession of the medium; and in her own energetic manner reminded him of her promise to come back if she could, and spoke a few sentences in her own well remembered voice. He decided therefore to invite Mrs Piper to his house in Liverpool. She left Lodge completely convinced of human survival, and of the power to communicate, under certain conditions, with those who had died.

After this demonstration he went to take part in an investigation of the powers of an Italian medium. Most of the spectacular powers of psychical phenomena have been demonstrated by mediums. There are two kinds of

2

32   *The Story of Electricity*

these, mental and physical. The mental mediums claim to be able to transmit messages from the minds of dead people. Here the medium in a trance passes the message by automatic writing or by speech. Physical mediums, on the other hand, produce levitation or the moving of objects in ways which contravene the laws of physics. William Crookes, who was knighted, and received the Order of Merit, was a believer in spiritualism, and supported Lodge. After Lodge returned to England he recorded, "I went in a state of scepticism as to the reality of physical movement produced without apparent contact, but this scepticism has been overcome by facts."

In 1896 Lodge gave a lecture on X-rays to the Liverpool Physical Society. He was one of the very first to provide X-ray photographs to help in diagnosis and surgery. Two years later he was awarded the Rumford Medal by the Royal Society. This is given every two years for the best work in heat or light. Many of his friends were pleased by the award, because they thought it went some way to redressing the injustice of the general public acclaim for wireless being given to Marconi, when it belonged more properly to Lodge. In the previous year he had written in 1897: "The only important discovery about the matter was made by Hertz in 1888. The receiver depends on cohesion under electrical influence which was noticed long ago by Lord Raleigh, and has been re-observed in other forms by other experimenters, including the writer in 1890."

On 19th May 1897 Lodge filed his first wireless patent. This was before Marconi's first patent was published, though—that was filed in the previous year. Lodge's patent was concerned with tuning, which allowed different signal frequencies to be separated, and also magnified the signal strength through the use of resonance. It was later to be the subject of a court case; at the time Lodge applied for an extension of its validity from 1911 to 1918. The extension was granted, though a concurrent agreement was reached with the Marconi Company that the Lodge-Muirhead Syndicate would then cease operations.

Two years later in 1899 the Marconi Company signalled across the English Channel from Dover to Wimereux. This success was a purely experimental result, since Marconi himself had no theoretical background for assuming if it might be possible. It must have been after this result that Marconi considered sending signals over longer sea paths, the next of which was to Crookhaven in the south-west of Ireland. When this in turn was successful it was a natural thought to consider sending signals across the whole width of the Atlantic Ocean. Here he had a problem in having to persuade the Directors of the Marconi Company to invest the large capital needed to build two large wireless stations in Cornwall and Newfoundland.

Because this was new engineering it almost inevitably meant that the designs were inadequate for the purpose. The result was that in stormy weather both expensive aerial structures collapsed. The responsibility then fell on Marconi to make the next decision, and almost inevitably he decided to make do with a makeshift transmitting aerial at Poldhu, and to use balloons and kites to carry an aerial wire as high as possible for receiving the signal in Newfoundland. The result is well known. He was able, in December 1901, to hear the agreed three dots of the letter S transmitted from Poldhu. It was natural that the news should create an immediate sensation in the popular press. But the scientific community, including Lodge, believed that the results were too few and too insubstantial to justify their immediate acceptance, and he wrote to *The Times* to make this point. But by this time Lodge had moved from Liverpool to Birmingham University.

Oliver Lodge and Birmingham University

Before Oliver Lodge accepted the offer of being Principal of Birmingham University he laid down three conditions. These were:

(1) He must be given his own research laboratory together with the funds to run it.

(2) Two of his scientific research assistants from Liverpool and his secretary must come with him.

(3) He was not to be debarred from psychical research.

It was clear to Lodge that an old university like Oxford, should specialise in, for example, archaeology and ancient philosophy, and be hesitant about letting old bottles be endangered by the inclusion of new wine. He wanted a broad education for the graduates of the new universities. They should be trained for life, and the schools should train for entry to the universities.

In 1902 Lodge was knighted. This was the same year in which King Edward VII instituted the Order of Merit on the occasion of his coronation. Although there are now 24 members, all of whom have gained distinction in military service or in the fields of science, art or literature, originally there were only 12 members. Of these 2 were members of parliament, 3 were soldiers, 2 sailors, 1 was an artist, and 4 were scientists. The scientists were Rayleigh, Kelvin, Huggins and Lister.

For Lodge the scientific outlook and his belief in the spirit world were complementary. It has been suggested that "Psychical phenomena were natural targets for the ambitious physicist who had already achieved success in explaining the mystery of electromagnetic waves." Lodge believed that the ether was the bond between the two. Myers and Lodge had sittings with Mrs Thompson, a London medium. Lodge has recalled hearing George Eliot, whom Myers had known well before her death, conversing with him through Mrs Thompson's control "Nellie".

Cross-correspondence experiments were instituted between 1905 and 1908. They involved two mediums who provided scripts, each in themselves without significance, but which yielded a significant message when brought together and compared. There were three groups involved in such an experiment:

(1) The automatist who provided the script. (In one case this was a Lecturer in Classics at Cambridge.)
(2) The interpreters who looked for the correspondence. (This could be Lodge.)
(3) The communicators who were the dead. (These included Myers and Mary Lyttelton. She was to have been the fiancée of Arthur Balfour, who was Prime Minister from 1902-1905, but she died on Palm Sunday 1875. He never married.)

The belief was that if the Myers intelligence survived his death it would make some effort to communicate in such a way as to eliminate the objections raised against previous experiments. After the experiment had been running for 7 or 8 years, Lodge wrote his tentative conclusions: "We find deceased friends constantly purporting to communicate, with the express purpose of patiently proving their identity, and giving us cross-correspondences between different mediums. Not easily do we make this admission." But the key figure in the cross-correspondences was Myers, who was believed by those involved to be the key figure "on the other side" in devising and arranging the experiment. After the death of Myers the bulk of the information communicated was by means of classical allusions, which were believed to be beyond the knowledge of those writing the scripts, but which would have been well-known to Myers.

Between 1907 and 1909 Lodge transferred his writings into the religious field. First he wrote, *The Substance of Faith Allied with Science: A Catechism for Parents and Teachers*. It consists of twenty questions and answers. This book went through six editions in the first three months. Two examples from it follow:

Q.1 "What are you?"

A.1 "I am a being alive and conscious upon this earth; a descendant of ancestors who rose by gradual processes from lower forms of animal life, and with struggle and suffering became man."

Q. 18. "What do you understand by prayer?"

A.18 "I understand that when our spirits are attuned to the Spirit of Righteousness, our hopes and aspirations exert an influence far beyond their conscious range, and in a true sense bring us into communication, with our Heavenly Father. This power of filial communication is called prayer; it is an attitude of mingled worship and supplication; we offer petitions in a spirit of trust and submission, and endeavour to realise the divine attributes, with the help and example of Christ."

There were four periods in Lodge's life:

(1) Up to age twenty-three when he first went to U.C.L.
(2) Up to age fifty during which he was a Professor in Liverpool for twenty years, and in which he developed skills as a lecturer and developed an understanding of electromagnetic waves.
(3) During the third period from 1900-1920 he was Principal of Birmingham University and he set up the Lodge-Muirhead Syndicate which for a number of years became a rival to the Marconi Company.
(4) The fourth period was that of his "retiral" from age seventy. During it he published ten more books between 1922 and 1927.

When in 1930 the *Spectator* invited its readers to compile a list of Britain's best brains, the top four were, Shaw, Lodge, Birkenhead and Churchill.

Lodge was a scientist and a Christian. He wrote to a Bishop in 1933: "I don't think any religion can be called scientific. The essence of Christianity lies in a recognition of the reality of the spiritual world, which was essentially and always in Christ's mind, and the information he gave about that world is not much, but is sufficient, namely that it is presided over by a loving Father. If that is true, and I doubt it not, then all that we have learned from science about the universe is subsidiary."

Soon after their Golden wedding in February 1929 Lady Lodge died "leaving him, as he believed, only temporarily separated from her, and that not completely, because he was connected through mediums with her and their son Raymond who had welcomed her on the other side."

Of his long interest in spiritualism Lord Rayleigh wrote after Lodge's death, "He had a reputation to lose, and he was prepared to risk losing it for the sake of what he believed to be the truth. How many of his critics would dare to do the same?"

Guglielmo Marconi (1875-1937)

His father was Guiseppa Marconi, a landowner, educated by priests, who owned an ancient house—Villa Grifone at Pontechio, eleven miles from Bologna, the city of Europe's oldest university. On this estate silkworms were bred and agriculture followed.

His mother was Anne Jameson, from County Wexford in Ireland, one of four daughters of Andrew Jameson, a Scotsman who had emigrated to Ireland and had established a distillery at Fairfield. They lived in Daphne Castle near Enniscorthy in Wexford, about sixty miles from Dublin. Annie Jameson had a wonderful singing voice who had been offered an engagement to sing at Covent Garden, but her parents thought that a career in opera was unthinkable for her. In compensation for this disappointment they offered her the opportunity to study *bel canto,* and this explained her visit to Italy.

The Jamesons had business dealings with a banking family in Bologna, with whom Annie stayed. She met there the son-in-law of the family, Guiseppe Marconi, a widower, seventeen years older than Annie. They fell in love, but the Jamesons refused consent. But when Annie reached the age of twenty-one she eloped and the couple were married in Boulogne. They then proceeded to Bologna.

After about a year their first child was born, to whom they gave the name Alfonso. Then there was a gap of nine years before Guglielmo became his brother. Guglielmo loved nature, and also fishing when he was young. Annie was determined that her sons should be bilingual, and instructed them herself in English, while using a tutor to teach Italian. She also had them brought up as Anglicans. After supper each evening she called the boys to her room and read them two chapters from the King James Version of the Bible. As their English improved they then read to her, and attended English Church Services in Florence and Leghorn. The bond that developed between Annie and Guglielmo was intense, but relations with his father were stiff by comparison, and he developed a reserve with most people, which he kept all his life. But when he was alone he demonstrated his manual skills, as when he took his cousin's sewing machine apart and then put it together again; or when he succeeded in distilling spirits at age thirteen, without getting caught. And then again he learned Morse code from an old telegraphist who was going blind, and to whom he used to read aloud to.

On the first of January 1894 Heinrich Hertz had died, and Guglielmo had read in a scientific journal an obituary written by Professor Righi of Bologna University. Although Guglielmo had failed to obtain entrance to Bologna University, his mother Annie persuaded Professor Righi to lend her son pieces of equipment to help him in his experiments. And Professor Righi also allowed him to sit in on his lectures and listened to Guglielmo's proposals. But the dedication was entirely that of Guglielmo's. He worked long hours attempting to improve both his transmitter and receiver. It has already been described how by chance he found that raising the height of his transmitting antenna he could extend the range of the received signal. On the other hand it was through many experiments that he found that the best metals to use in his detector were nickel and silver, with 95% of the first and 5% of the second in the coherer. And that evacuating the air from the tube provided a further very significant improvement. Then when Guglielmo discovered that they could detect a signal which had come over a hill it was necessary to look for Government support.

Because the Italian Navy was building up its fleet it may have been that the Minister for the Navy might have supported Guglielmo's work. But the application was sent to the Minister of Post and Telegraph who rejected the opportunity of supporting the work. At this point his mother Annie believed that since the invention could have application to ship-to-shore communications they should take it to England, the greatest maritime nation in the world. In later years Guglielmo said, "Mind you, Italy did not say the invention was worthless, but wireless in those days seemed to hold promise for the sea, so off to England I went."

In February 1896 Annie and Guglielmo set out for England. By way of family preparation they were to be met by a cousin Henry Jameson-Davis at Victoria station. He himself was an engineer, and in a short time he was able to take Guglielmo to a good patent lawyer in London. Nevertheless it took four months to prepare the patent papers, which on 2nd June were deposited as a provisional specification in the London Patent office. This was followed on 2nd March 1897 by the complete specification, which was accepted as Patent No.12039 on 2nd July 1897.

In addition to having the protection of a patent, the family were advised to obtain a favourable opinion from an engineer of substance. W. H. Preece the chief engineer at the Post Office came into this category, and an introduction was arranged through a friend of Henry Jameson-Davis. Preece's professional interest arose from having the desire to provide lightship keepers with early information on approaching storms at sea. The meeting with Guglielmo was a success, and in the first demonstration a range of half a mile was achieved, even after the signal went through intervening masonry walls. In following experiments near Stonehenge a range of nine miles was achieved.

A bank in Milan then offered 300,000 lire for rights to his invention, an offer which his father Giuseppe recommended he should accept. But Guglielmo refused, believing it was worth more than that. He also had the problem of doing three years of military service in Italy. This was got over eventually through the King giving him permission to stay in England as a naval cadet in training with the Italian Embassy in London. The rank associated with this naval cadetship was that of midshipman, and as long as he held this it was paid anonymously into the account of the Italian Hospital in London. Throughout his life Marconi regarded himself as a scientific vessel, a human instrument, chosen by a higher power to make a unique contribution to the progress of mankind. For his great discoveries he credited the Divine Will.

On 26th June 1900 Marconi took out a provisional patent on selective tuning to allow one particular set out of several receiving stations to receive a signal. This was taken out in the U.S.A. The full patent was granted later with the number 7777. In the same year Poldhu was chosen as the site of the U.K. end of the transatlantic link to America which it was his ambition to achieve. But as stated previously, as a careful engineer he started off by sending a signal first to Crookhaven, at the south-west tip of Ireland, about sixty miles from Cork. The range for this transmission was 225 miles, about one tenth of the distance across the whole Atlantic. The design of the transmitting station at Poldhu was undertaken by Professor J.A. Fleming of University College, London. By September the designed structure with two hundred-foot high supports was ready to support the conical aerial to be used at the transmitting end, but a cyclone blew everything down before it could be attached.

For financial reasons Marconi then chose to support a simple wire aerial of total weight ten pounds, by means of a balloon fourteen feet in diameter filled with one thousand cubic feet of hydrogen gas. They had two balloons available, but one was lost when in a gale they decided to fasten it to the ground more securely, and in doing so the rope broke. Marconi then decided to use kites for supporting the wire transmitting aerial. A kite rose to a height of four hundred feet, which was better than the height achieved by the balloon, but its position was significantly more variable in a strong wind.

The site in Newfoundland where the receiving aerial was placed was at Cape Cod. Again a storm blew down the supports of the original design of aerial, and resort was made once again to the use of a kite for supporting the aerial wire. On 12th December 1901 the three dots of the letter S in Morse code were heard, transmitted from Poldhu. Marconi was then aged twenty-seven. It had taken four years since the patent had been granted, and he had substantial help from William Preece of the Post Office with whom he had established good relations. His own experimental skills had also contributed to the building up of a team of reliable assistants. But the financial success of a company demands more than technical success. And he had opposition from the cable companies which stood to lose money if wireless became profitable.

But about this time Marconi fell in love with an Irish girl, nineteen year old Beatrice O'Brien, daughter of the thirteenth Baron Inchiquin. The Baron kept a residence in London and sat, while Parliament was in session, as a representative peer for Ireland. He had thirteen children, of whom seven were girls and six were boys. Marconi married Beatrice on 16th March 1905 in St George's, Hanover Square, London. In 1908 their second daughter Degna Marconi was born, who has written a perceptive biography of her parents. The following year Marconi was awarded the Nobel Prize for Physics, shared with Professor Ferdinand Braun of the Physical Institute in Strasbourg. When the two prize-winners met Professor Braun said he thought the whole prize should have gone to Marconi.

Marconi continued with his experiments at sea, and in 1910 was able to receive signals which had travelled from Clifden in Ireland 4000 miles in daytime, and 6735 miles at night. The Titanic disaster in 1912 would have been worse in the absence of wireless telegraphy. So great was the appreciation of those who survived that they marched to Marconi's hotel and presented him with a gold medal. In 1913 Marconi was created a Senator in the Italian Parliament. This is a post for life and is restricted to those who are aged over forty years of age. Nevertheless he continued his experimental researches, particularly with short waves and succeeded effectively in spanning the whole globe.

CHAPTER NINE

PLANCK : PERSONAL COMMUNION WITH GOD

FEYNMAN : PERSONAL COMMUNICATION WITH MAN

(c.1900 – c.1950)

FOLLOWING ON FROM CHAPTER EIGHT with its description of the achievements of Hertz, Heaviside, Marconi and Lodge, we come to the twentieth century in which wave theory is broadened to include waves of matter in the form of charged particles such as electrons and atoms. This was necessary because Maxwell's equations were found not to apply to the small distances associated with atoms, and even more so to the still smaller size of the nucleus. The advances made by including waves of matter were produced both by mathematicians such as Hermann Weyl (1885-1955), born in Germany, whose work was to flower in both Switzerland and at Princeton, and Erwin Shrodinger (1887-1961). He was an Austrian who found political refuge first in Germany, then further in Switzerland, then back to Germany, Oxford, and finally to Dublin. In these varied countries he held different University appointments.

In addition to the work of the mathematicians, much credit is due to theoretical physicists as well. Of the two selected for inclusion in this chapter the first is Paul Dirac (1902-1984), born in Bristol of a Swiss father and an English mother. Because of his father's insistence that conversation in the dining room should be in the French language alone, Paul's rebellion against this took the form of not speaking there at all. This absence of spoken communication then extended to the whole, not only of his boyhood, but to the whole of his life. It was a result compounded by the fact that at school he was taught never to start a sentence without knowing how he was going to

241

end it. The result, in culmination of this practice, was that his research students in Cambridge coined the word "dirac" as a unit for speaking at the rate of one word per year.

The second theoretical physicist to be selected is Richard Feynman (1918-1988). In his Nobel Lecture of 1965 he records that at age nineteen, while still an undergraduate at the Massachusetts Institute of Technology, he was reading Dirac's *The Principles of Quantum Mechanics*, and that the last sentence in that book became imprinted in his memory—"It seems that some essentially new physical ideas are here required"—to explain the quantum theory of electricity and magnetism. After graduation from M.I.T. in 1939 Richard Feynman moved to Princeton University to begin his Ph.D studies under the supervision of John Wheeler. It will be recalled that at the Princeton Institute of Advanced Studies, Einstein had been given a permanent post, and he chose to attend Feynman's first Seminar. After the requisite period of attendance Feynman's doctorate was awarded in 1942, but before this time he had worked on the atomic bomb Project at Princeton from 1941 to 1942, and then later at Los Alomos from 1943 to 1945. Following the end of World War II he returned to University life, first at Cornell, and then from 1950 at the California Institute of Technology.

Of these four twentieth-century contributors three were Nobel Prize-winners, Shrodinger and Dirac in 1933, and Feynman in 1965. Because Weyl was undoubtedly a mathematician and there is no Nobel Prize in Mathematics, he has been included because of his intellectual stature among his peers. Weyl made major contributions in every field he entered, and is to be compared with the great mathematicians David Hilbert and Henri Poincare.

But in listing these four twentieth-century contributors alone, one is conscious of apparently undervaluing many other famous physicists. Among these the names of M. K. E. L. Planck, Niels Henrik, David Bohr, Lois Victor de Broglie, Wolfgang Pauli, Werner Karl Heisenberg and Hideki Yukawa, all of whom were themselves recipients of Nobel Prizes, come to mind. To obviate any charge of unfairness against these Nobel Laureates it has therefore been considered desirable to include a short biographical sketch of each of them, along with Tomonaga and Schwinger who shared the 1965 Nobel Prize with Feynman. An attempt has thus been made to describe their significant contributions to the history of electricity. This will then be followed by the initial four twentieth-century contributors, Weyl, Schrodinger, Dirac and Feynman in turn.

M. K. E. L. Planck (23/4/ 1858—3/10/1947)

Max Karl Ernst Ludwig Planck was the fourth child of Johann Julius Wilhelm Planck who was Professor of Jurisprudence at Kiel University. His great-grandfather had been Professor of Theology at Gottingen, holding what have been described as "Enlightenment views of rationalism and tolerance; with God, not Christ at the centre of belief". In turn his son also became a professor of Theology at Gottingen, before his son Johann Julius Wilhelm Planck became the Professor of Jurisprudence at Kiel. Wlhelm Planck remained at Kiel until 1867 before moving to Munich where he died in 1900.

Max Planck then received his elementary education at Kiel, but moved into a classical gymnasium in Munich at age nine. It was there he learned from his mathematics teacher Hermann Muller, the first scientific law which possessed universal validity, namely the principle of the conservation of energy. Planck, in his Scientific Autobiography, tells us of how Muller "at his raconteur's best, tells the story of the bricklayer lifting with great effort a heavy block of stone to the roof of a house. The work he thus performs does not get lost; it remains stored up, perhaps for many years, undiminished and latent in the block of stone, until one day the block is perhaps loosened and drops on the head of some passerby."

Much of what follows depends on further information obtained from Planck's Scientific Autobiography.

After secondary school Planck entered the University of Munich at age sixteen. He initially decided to study Mathematics and experimental Physics, though other options such as the study of language, and also music were considered. There were no classes in Theoretical Physics at the time, so this choice of experimental Physics provided the only time in Planck's life when he had the opportunity of actually carrying out experiments.

After three years at Munich Planck spent the academic year 1877-1878 in Berlin. There he attended lectures by Kirchhoff and Helmholtz. His view on Helmholtz was that his lectures were not prepared properly, and that the class bored him. Eventually, his classes became more and more deserted, and finally they were attended only by three students.

Kirchhoff was the very opposite. He would always deliver a carefully prepared lecture, with every phrase well balanced and in its proper place. Not a word too few, not one too many. But it would sound like a memorized text, dry and monotonous. We would admire him, but not what he was saying.

Under these circumstances it was necessary to do his own reading on subjects which interested him, subjects in particular relating to the energy principle. One day he came across the treatises of Rudolf Clausius, whose lucid style and clarity of reasoning made an enormous impression on him. He appreciated especially his exact formulation of the two Laws of Thermodynamics. (The first being the Principle of the Conservation of Energy.) With regard to the Second Law, up to that time, as a consequence of the theory that heat is a substance, the universally accepted view had been that the passing of heat from a higher to a lower temperature was analogous to the sinking of a weight from a higher to a lower position, and it was not easy to overcome this mistaken opinion.

Clausius deduced his proof of the Second Law of Thermodynamics from the hypothesis that "heat will not pass spontaneously from a colder to a hotter body". But this hypothesis must be supplemented by a clarifying explanation. For it is meant to express not only that heat will not pass directly from a colder to a warmer body, but also that it is impossible to transmit by any means, heat from a colder to a hotter body without there remaining in nature some change to serve as compensation.

To express this more simply, "The process of heat conduction cannot be completely reversed by any means." Whether a process is reversible or irreversible depends solely on the initial state and the terminal state of the process. In the case of an irreversible process the terminal process is in a certain sense more important than the initial state—as if, so to speak, Nature "preferred" it to the latter. He saw a measure of this "preference" in Clausius's entropy; and he found the meaning of the Second Law of Thermodynamics in the principle that in every natural process the sum of the entropies of all bodies involved in the process increases. He worked out these ideas in his doctoral dissertation at the University of Munich, which he completed in 1879.

Planck then goes on to say that the effect of his dissertation on the physicists of those days was nil. He believed Helmholtz probably did not read his paper and Kirchhoff expressly disapproved of its contents. As for Clausius, he did not answer his letters and when Planck tried to see him at home in Bonn he was not in. But Planck carried on his investigations on entropy, writing a number of monographs within the general title of *On the Principle of the Increase of Entropy*. He was later to find that his results had been previously obtained by J. W. Gibbs in America.

In 1885 at age twenty-seven he was offered an Associate Professor's post at the University of Kiel, where his father still held the Chair of Jurisprudence. This also enabled him to marry his fiancée Marie Merck from Munich. Four years later he accepted a similar post in Berlin after the death of Kirchhoff, and after a further three years he was promoted to a full Chair there. It was in Berlin that he remained until his retirement in 1928 at the age of seventy. During his time in Berlin it was natural that he should interact with Helmholtz who was the world authority on acoustics. In contrast to Helmholtz who preferred the tempered scale, Planck preferred the natural scale. Soon after his arrival in Berlin the Institute of Theoretical Studies received a large Harmonium of untempered tuning, built by the Schiedmayer piano factory in Stuttgart. The construction delivered one hundred and four tones in each octave. In his study of this instrument he discovered that the tempered scale was more pleasing to the human ear, under all circumstances, than the untempered scale. Before the time of J. S. Bach the tempered scale had not been universally known.

In 1900 Planck introduced the quantum theory to solve the problem of black body radiation. He began by considering an irreversible process. This is defined as referring to an initial state A and ending with a final state B, such that no return is possible by any means to the initial state A again. This represents the practical situation in any thermodynamic process, which always results in an increase of entropy. His result showed that the energy emitted by an oscillator could take on only discrete values or quanta. The quantum of energy for an oscillator of frequency is hf, where h is Planck's constant 6.624 times 10^{-34} *joule*-secs. Because it represented the product of energy times time, Planck called it the elementary quantum of action. And showing his own anxiety at the implications of his analysis, he continued, "Although it was absolutely indispensable for obtaining the correct expression for the entropy—it proved itself cumbersome and unmanageable in all attempts to make it fit in any suitable form into the framework of the classical theory.—The failure of all attempts to bridge this gap left no doubt that either the quantum of action was merely a fictitious quantity—in which case the entire deduction of the radiation formula was illusory in principle and represented nothing more than empty juggling with symbols—or else the deduction of the radiation formula rested on an actual physical fact, in which case the quantum of action must play a fundamental part in physics and appears as something quite new and hitherto unheard of, compelling us to recast our ideas from the very bottom—ideas which from the invention of the infinitesimal calculus by Leibniz and Newton rested on the assumption of continuity in all causal relations. Experiment has decided in favour of the second alternative, giving energy something of an atomic nature, each of the infinite number of possible frequencies "f" having associated with it a definite quantity of energy "hf", so that any transfer of energy at the frequency "f" took place only in integral multiples of "hf". The tiny wavelengths of the matter waves precluded experimental proof of their existence till later.

Max Planck as a Preacher

Considering that Max Planck's great-grandfather and his grandfather were both Professors of Theology, it is perhaps not too surprising that he himself should have taken up the role of preaching after his scientific career had ended. He had been raised as a Lutheran, and from 1920 until his death in 1947 he had been an elder in his congregation in Berlin-Grunewald. Moreover, he was accustomed to saying grace at meal times, and it is recorded that he was shocked at a friend's scepticism about religion.

It was in a lecture entitled "Religion and Natural Science" given in 1937, that he delivered a carefully thought out distillation of his views on the unity of knowledge, covering in particular the supposed difficulty of a scientist believing in religion.

In this lecture Planck begins by quoting from Goethe's *Faust* the words of Marguerite, the principal female character of Part I, a girl of humble station, simple, confiding and affectionate. "She asks Faust 'Tell me—how do you stand to religion?' It was the worried question of an innocent girl, in fear for her new found happiness, to her lover whom she recognises as a higher authority."

But the question she asked is of universal relevance. Inside every thinking mind there is a desire for peace of mind based on all the knowledge it possesses. This desire may or may not reach a conclusion. If it does, strength is given to that mind. If it does not, embarrassment follows to the person of whom the question is asked. And that is what happens to Faust. He gives a defensive reply: "I want to deprive nobody of his sentiments and his church."

Faust is being honest. He is unable to say "I believe all the doctrines of the Church", because that would mean saying "I recognise as truth all the doctrines of the Church". And this raises the question "Of which Church?" But when we apply Marguerite's question to ourselves, it may be helpful to bear in mind that at that time, Faust was under the power of Mephistopheles. And so it is necessary to consider the existence of forces, both good and evil, which may operate on us too.

Planck defines Religion as *the link which binds man to God.* It is based on our humility before Him. But there is also an active desire on our part to be in harmony with Him, and to be protected from visible and invisible dangers which surround. The result of this is to produce an inner peace of mind and soul, and a trusting faith in His omnipotence and benevolence.

But the significance of religion goes beyond the individual. It applies also to the larger community, to a congregation, to a country and even to the whole world. With regard to the whole world, in religion God is the starting point, whereas in science He is the crown of the structure. Science wants man to learn, but religion wants him to act. For this religious acting there must be a direct link to God. "With this link there comes an inner firmness and peace of mind which is the highest boon of life."

Planck's rallying cry for both religion and science is, "On to God!"

Niels Henrik David Bohr (7/10/1885—18/11/1962)

The religious outlook of Planck may be said to have controlled his life. But this is not so for the physicist Niels Bohr, whom we now go on to consider. Niels Bohr was the son of a University Lecturer in Physiology at the University of Copenhagen. His mother was the daughter of a Jewish politician who owned a stately home opposite the castle where the Danish parliament sat. Niels was born in this home. A few months later his family moved into a Professorial house in Copenhagen when his father had been appointed to a University post in the city.

As a child Niels Bohr did not receive religious instruction. He was, however, a gifted reader, and as a boy enjoyed the Icelandic Sagas. It was from one of these Sagas that he used to quote the saying, "He went out to gather together words and thoughts." He was familiar too with the writings of Soren Kiekegaard, and admired them linguistically, rather than from the religious viewpoint from which they were written. With regard to his ability in foreign languages it is said that he read *The Pickwick Papers* with a dictionary in his hand, he liked Othello in Shakespeare and he also loved P. G. Wodehouse. His German was good, being fond of Schiller, but his French was poor. And his reputation as a public speaker was reputed to be divinely bad, though he himself did not share this view.

As a child Niels and both his siblings were baptised at the same time, when Niels was aged six. It appears that Niels' father would take him to church on Christmas Eve, just so that Niels would not feel different from other boys. Similarly, with regard to christening, the parents did not want their children to be the only children in their class who had not been baptised.

When it came to his wedding on 1ˢᵗ August 1912 the decision had been taken that it was to be a civil ceremony. In confirmation of this Niels had resigned his membership of the Lutheran Church some months before. In contrast to this his bride was still nominally a member of the Lutheran church at that time. Their civil ceremony paralleled that of Niels' parents, and thus explains the absence of a religious upbringing of their children.

The final comment in this section comes from a story told by Abraham Pais in his biography *Niels Bohr's Times*. When Elizabeth Bergner asked Einstein whether he believed in God, he replied, "One may not ask that of someone who with growing amazement attempts to explore and understand the authoritative order in the universe." When she asked why not, he answered, "Because he would probably break down when faced with such a question." Pais then goes on: "Imagery like that would never occur to Bohr's mind."

Niel Bohr's Growth in Physics

In September 1911 Niels Bohr was on his way to Cambridge, financed by the Carlsberg Foundation in Denmark. He had in the preceding May, obtained his D.Phil. for a thesis dealing with the electron theory of metals. After he had obtained this degree he had a personal copy bound with blank sheets between the pages on which he marked corrections, addenda and deletions. It has been remarked that this was a typical characteristic of Niels Bohr. The choice of Cambridge rather than Berlin was probably because of the historic sequence of Maxwell, Rayleigh and now J. J. Thomson who controlled its workings. Unfortunately the first meeting between Bohr and Thomson was marred by Bohr bringing a copy of Thomson's book to that meeting, pointing to a certain page and saying politely, "This is wrong." Thomson was a theoretician who had been awarded the Nobel Prize for Physics in 1906, and almost certainly took unkindly to this criticism. His strength theoretically was not matched by experimental skills in which he has been described as handless. Indeed it has been suggested by another Nobel Laureate, Sir Edward Appleton, that the Nobel Prize he was given might more fairly have been awarded to J. J. Thomson and his Chief Technician. The result of the dislike between Bohr and Thomson was that Bohr transferred at the end of 1911 to Manchester, where Rutherford was Professor.

In 1911 Rutherford in Manchester suggested that the atom consisted of a small positively charged nucleus round which a number of electrons revolved to keep the atom electrically neutral. This suggestion was based on his experimental results which showed that positive alpha particles sometimes showed a sharp deflection from a straight line path, due to their passing through the strong electric field surrounding the central charge.

In 1913 Bohr, working now in Rutherford's laboratory, modified Rutherford's theory. It was known that light radiating even from the simplest Hydrogen gas consisted of a spectrum of a number of separate lines. Hence Bohr postulated that this spectrum resulted from a jump on the part of the electron from one of a small number of possible orbits to another in which the electron possessed less energy. The difference in these energies was equal to the frequency of the spectrum line multiplied by Planck's constant "h". The experimental confirmation of this was already available from results by J. J. Balmer in 1885. Bohr's calculations of assumed circular orbits for the electron paths were later extended by Arnold Sommerfeld to elliptical paths. This provided an explanation for the fine structure shown by certain spectral lines.

Thus Bohr's hypothesis made him a world figure in science. The respective energies associated with the discrete orbits of an electron inside the hydrogen atom he denoted by E_n, where n is equal to 1, 2, 3 etc. The value for n equal to 1 corresponds to what he called the ground state, meaning the orbit closest to the nucleus. In terms of classical theory this electron would spiral into the nucleus. To get over this difficulty Bohr simply postulated that the ground state was stable, and did not follow conventional classical theory. But he also drew attention to the role of the outermost ring of electrons as being associated with the chemical properties of the elements, i.e. their valencies.

In 1914 Bohr applied to the Government Department of Religious and Educational affairs asking for the establishment of a Chair in Theoretical Physics, expecting that he would be appointed to that post should the Chair be granted. Rutherford was asked to support the establishment of this Chair, but in addition offered Bohr a readership in Manchester. Because of administrative delays in Denmark, Bohr accepted this Readership, and because of the outbreak of war succeeded in reaching Manchester only by travelling round the North coast of Scotland. He and his wife stayed in Manchester until 1916, when they were able to return to Copenhagen to the much delayed Chair.

For five years after his return Bohr was much occupied with the design and supervision of his new Institute. This plus the giving of lectures was responsible for an eventual breakdown. As a result it appears that after 1921 he gave up lecturing to undergraduates.

In the following year Bohr was awarded the Nobel prize "For his studies on the structure of atoms and the radiation emanating from them". In 1923 de Broglie associated waves with electrons, and with material particles also. Bohr also proposed that uranium fission by slow neutrons is due to the presence of the rare isotope 235.

Louis Victor Pierre Raymond duc de Broglie (15/8/1892—19/3/1987)

Louis de Broglie's father was the Duc de Broglie. During his school career Louis developed an interest in history, and entered the Sorbonne with the intention of making a career in the Diplomatic Service. After being assigned a research topic in history he decided instead to study for a degree in theoretical physics, and was awarded a Licenciate in Science in 1913. During the First World War he served in the army, attached to a wireless telegraphy section which operated from the Eiffel Tower. On resuming his studies in 1920 he was captivated by the work of Planck who in 1900 had proposed that electromagnetic energy arrived in bundles of magnitude "hf", where h is Planck's constant and f is the frequency. This subject then formed the substance of his doctoral thesis *Researches sur la theorie des quanta* which was submitted in 1924. This proposed the particle-wave duality theory, that matter has the properties of both waves and particles.

Although all theses are expected to demonstrate novelty in some respect, it will be understood that the extreme novelty in the single substantive idea of this thesis, put an above average responsibility on the examiners. So much was this the case that the thesis supervisor M. Langevin asked for an additional copy of the thesis to be sent to Einstein to ask for his opinion. The reply from Einstein was that the central idea seemed quite interesting to him. On the basis of this reply the thesis was accepted by M. Langevin.

At that time the idea of Bohr's electronic orbits was becoming accepted. That is to say in the simplest case of a hydrogen atom, Bohr had proposed that instead of the orbiting single electron collapsing into the positive nucleus, it formed a stable circular orbit round the nucleus. De Broglie now suggested that accompanying this circular path was a wave whose length was equal to the circumference of the first stable circular orbit. This then provided a measure of the radius of the atom, now known as the Bohr radius. To achieve this result it is necessary to equate the centrifugal force due to the orbital motion, to the electrical attraction between the electron and the nucleus. This works out at 0.53 Angstroms. But De Broglie also showed that the length of the wave was equal to Planck's constant divided by the product of the mass and velocity of the electron.

This was a purely theoretical result, but so significant that it appeared that if a beam of electrons in space were accompanied also by such pilot waves, diffraction phenomena should be obtained. Experiments to confirm this were carried out by G. P. Thomson in Aberdeen and C. J. Davisson in New York. The experiments were successful and both workers were awarded Nobel Prizes in 1937.

De Broglie's contribution was analogous to Einstein's who had ascribed particle-like properties to light. In De Broglies's case he ascribed wavelike properties to particles of matter. In the general case, there are waves of different radii for even the hydrogen atom, and these have different velocities of propagation. At regular intervals the waves will combine to form a crest. The velocity of the crest De Broglie identified with the velocity of the electron, and the distance between successive crests is the De Broglie wavelength. Even as a wave is characterised by its wavelength, so a particle is characterised by its momentum. De Broglie found that the wavelength of the matter wave and the momentum of the particle are related by the product of wavelength and momentum being equal to Planck's constant. In terms of a numerical example, for a slow electron moving at 100 cm per second, the De Broglie wavelength is about 0.07 cm, while for an alpha particle emitted by radium it is about 0.7 Angstrom, approximately the diameter of the atomic nucleus.

Wolfgang Pauli (25/4/1900-15/12 1958)

Wolfgang Pauli, born in Vienna, was the son of a medical doctor who later became a Professor. He was also a godson of Ernst Mach. In his high school years he had already demonstrated such marked ability that when he arrived at Munich University in 1918, to study under Arnold Sommerfeld, he brought with him a paper on general relativity. After further study this was published in German in 1921, eliciting from Einstein the comment, "Whoever studies this mature and grandly conceived paper might not believe that its author is a twenty-one year old man." The paper was translated into English and published in 1958, and has been described as "One of the best presentations of the subject".

In 1921 he received his Ph.D. summa cum laude from Munich University, and then moved to Gottingen for six months to work with Max Born. While he was there he met Niels Bohr and wrote, "A new phase of my scientific life began when I met him." He was invited to spend a year in Copenhagen. From 1923 to 1928 he was lecturer in the University of Hamburg, and then was appointed Professor at the Federal Institute of Technology in Zurich. After several Visiting Professor posts in the United States, he returned to Zurich where he died in 1958.

Wolfgang Pauli is remembered for his Pauli Principle (or Exclusion Principle). As a result it became possible to construct models for all atoms from hydrogen to uranium. It will be recalled that the hydrogen atom consists of a single electron revolving about the nucleus in any one of a number of circular orbits, with the numeric 1 being used for the orbit nearest the nucleus, and 2, 3 etc. for larger radii. These numbers are referred to as "Quantum numbers". The difference in energy levels between adjacent circles is "hf", where "h" is Planck's constant and "f" is the frequency of the light emitted by the electron in jumping from the higher orbit to the lower one. In the event of some orbits being elliptical rather than circular, as calculated by Sommerfeld, similar arguments apply. To ensure that the difference in energy can have certain values only, a second quantum number "k" is introduced which is equal to the ratio of the major to minor axis of the ellipse, with values between 1 and n. When "k" is equal to n the axes of the ellipse are equal and the ellipse becomes a circle.

A third quantum number is required to take account of the splitting of the lines in an external magnetic field—the Zeeman effect. The magnetic field produced by an electron describing an orbit causes a swaying motion of the plane of the orbit. Pauli showed that a line at right angles to the plane of the orbit, through the centre, describes a cone about the direction of the external magnetic field. The third quantum number "m" restricts the angle of this cone to certain discrete values.

Pauli found it necessary to introduce a fourth quantum number. This fitted in with the idea of electron spin with the two possible values of $+\frac{1}{2}$ and $-\frac{1}{2}$. A spinning electron possesses rotational angular momentum and also behaves as a small magnet because it is electrically charged.

The exclusion principle says: "No two electrons in the same atom can have all four numbers equal." To see how this decides the closing of the electronic shells, starting with the hydrogen atom in its ground state, i.e. with its single electron in its innermost orbit, gives n=1, m=0, m=0. For the helium atom with one more electron, n=1, k=1, m=0. Then by the exclusion principle their fourth quantum numbers must be different, so that one has $s=+\frac{1}{2}$ and the other $s=-\frac{1}{2}$.

Wolfgang Pauli's Neutrino

Most discoveries in applied physics are first made experimentally. In such cases the theory for them follows, after a period of time which depends on the complexity of the mathematics involved. It is therefore pleasing to report on the discovery of the neutrino by Wolfgang Pauli, that the theory for it was first made in 1911, but its experimental confirmation did not take place until 1956. The reason for this is that the experimental set up to demonstrate the effect made use of a nuclear reactor. In this reactor neutrinos escaped in very large numbers, with about ten million million crossing one square centimetre per second at sea level. But in this experiment only three neutrinos per hour were observed. Another experiment carried out in a very deep Gold Mine in South Dakota, to reduce the number of cosmic rays from outer space entering the apparatus, and using solar neutrinos gave a detection rate of only one third of what was expected.

It is helpful to go back in time to the beginning of nuclear physics to understand the difficulties faced by the investigators of that time. Radioactive elements emit Alpha waves, Beta waves and Gamma waves. The Alpha waves were known to be the nuclei of helium atoms. Beta waves were electrons which were sometimes emitted by nuclei following an Alpha wave emission to restore the balance of charge and mass upset by the Alpha emission. Gamma waves are short electromagnetic waves emitted by atoms which have radiated Alpha or Beta waves.

The Alpha waves or particles all have the same energy corresponding to the difference between the mother and daughter energies. Taking the Gamma rays next they have complex sharper lines than the lines of optical spectra. But the Beta particles do not have well defined energies. Their spectrum extends continuously from practically zero to large values. This suggested to Bohr that the Law of Conservation of Energy did not apply to Beta emission and he suggested that this might account for the energy of the sun.

But Wolfgang Pauli did not accept this. He took the alternative view that there might be another, as yet unknown, particle emitted along with the Beta particle. And he postulated that the sum of their and the Beta particles' energies was always the same. This would then guarantee Conservation of Energy as before.

Some understanding of the small number of neutrinos observed experimentally was forecast by Pauli in 1930 when he listed the following properties of neutrinos:

(1) They are electrically neutral, so that they are unaffected by electric charges in any detector, and their magnetic moment is zero.
(2) They satisfy the Exclusion Principle.
(3) They have a spin of ½
(4) They have a mass of the same order as an electron.

In making these predictions Pauli insisted that conservation of energy, momentum, and angular momentum were maintained. This was in opposition to the views of Niels Bohr at the time who had an alternative theory which did not maintain all these conservations.

In addition to the generation of neutrinos it is important to consider the effect of neutrons on the generation of further electrons. Neutrons are more massive than protons, and left to themselves will eventually decay to a proton, with a positive charge, and an electron with a negative charge. But momentum and energy will not be conserved unless another neutral particle is involved, and Pauli thus postulated the neutrino.

It sometimes happens that the originator of an idea benefits from those who work on it afterwards. This has happened in the case of Wolfgang Pauli. Also for E. Fermi who, in what has been described as his greatest discovery in theoretical physics, enunciated the following rule:

> "For every transition from a neutron to a proton there is associated a creation of an electron and a neutrino. For the inverse process from a proton to a neutron there is the disappearance of an electron and a neutrino."

Werner Carl Heisenberg (5/12/1901—1/2/1976)

Werner Heisenberg was born in Wurzburg in Bavaria in 1901. His father was a secondary schoolteacher of Greek, who progressed to become in Munich, Germany's only Professor of Greek. Werner grew up in Munich, showing himself at school to be an independent worker in physics, and also an accomplished piano player, at which he excelled throughout his life. He enrolled, at age nineteen, in the University of Munich to study physics under Arnold Somerfeld, and continued his studies at Gottingen until 1923, studying physics under Max Born and mathematics under David Hilbert. There was an important turning point in his studies in 1922 when Niels Bohr gave a lecture at Gottingen. At the end of the lecture Heisenberg got up to make an objection. It is reported that Bohr was hesitant in his reply, but invited Heisenberg to go on a walk with him that afternoon. The discussion lasted about three hours, and after it was over Bohr's comment to a friend was, "Heisenberg understands everything."

Heisenberg received his doctorate in 1923 from Munich, under Sommerfeld, on the topic of turbulence, discussing both the stability of laminar flow and the nature of turbulent flow. This required the solution of a fourth order linear differential equation. There is a story relating to this examination that during its course Heisenberg antagonised the External Examiner. Consequently when Heisenberg showed ignorance of some rather elementary optical information under general questioning, the External examiner decided to award a Pass mark at the lowest level only.

From 1924 to 1927 Heisenberg was a Research Fellow at Gottingen, but within this period he was given an International Rockefeller Foundation Fellowship to do research with Niels Bohr at Copenhagen. He then returned to Gottingen to work with Max Born, developing the matrix mechanics formulation of Quantum Mechanics. But in 1926 he was back in Copenhagen as lecturer and assistant to Niels Bohr, and in 1927developed his Imprecision Principle, commonly known as the Uncertainty Principle. In the same year he was appointed Professor of Theoretical Physics at Leipzig University, retaining that post until 1941.

The Uncertainty Principle says that it is not possible to know precisely the exact position and momentum of a particle. To know the exact position very accurately it is necessary to use a high frequency of light which implies a large energy from Planck's Law which says that the energy is of magnitude "hf" where "h" is Planck's constant. This energy will give a kick to the particle, disturbing its position. Similarly to know the momentum accurately we must give a small kick which implies a long wavelength and this implies a large uncertainty in position Thus Heisenberg went on to show that the product of the two uncertainties in position and momentum, dx. times dm, is equal to "h".

Returning now to Heisenberg's development of matrix mechanics he treated the atom as if its line frequencies coincided with all possible frequencies that the atom could emit. "His aim was to establish a basis for theoretical quantum mechanics, founded exclusively upon relationships between quantities which are in principle observable—which atomic orbits are not." Since he worked with matrices for which the product is not commutative he was able to show that the difference of the products is equal to h/(2 pi j), where "j" is the square root of minus one.

Hideki Yukawa (23/1/1907—8/9/1981)

Hideki Yukawa was born in Tokyo in 1907, a son of Takugi Ogawa, who later became Professor of Geology at Kyoto University. Hideki Yukawa was brought up in Kyoto and graduated there in 1929.

It is of interest to examine the way in which physics research developed in Japan up to the time in which Hideki Yukawa became a Nobel Laureate in

1949. Going back to the 1880's we come across Hantaro Nagaoka who graduated from the University of Tokyo in 1887. Between 1892 and 1896 Nagaoka studied at the Universities of Vienna, Berlin and Munich After his return to Japan he became Professor of Physics at the University of Tokyo, from 1901 to 1925, where his pupils included Hideki Yukawa.

In 1904 he developed a partly correct model of the atom, which was based on an analogy to the stability of Saturn's rings. His correct predictions were that the atom consisted of a very massive nucleus, in the analogy to the very massive planet. Secondly that electrons revolved around the nucleus, bound by electrostatic forces, in analogy to the rings revolving round Saturn, bound by gravitational forces. Another of his ideas which still finds use is in the factor of Nagaoka's Constant, which enables the inductance of a single layer solenoid to be evaluated from its dimensions.

While Nagaoka was at Tokyo University he was responsible for sending Nishina for over five years to Niels Bohr's Institute in Copenhagen, and several other young Japanese scientists to the same Institute in the 1920's. During the same period reciprocal visits were arranged by him which enabled Dirac, Einstein, Heisenberg and Sommerfeld to visit Japan.

After Nishina returned to Japan following his five years in Copenhagen, he became the founder of experimental and cosmic ray research in his home country. As part of this responsibility he supervised the construction of several accelerators in Tokyo. The most advanced of these was a 150 cm. cyclotron which was destroyed by American Occupation forces after the war.

When Osaka University was founded in 1931 Nagaoka became its first President. In 1933 one of the appointments he made was that of Hideki Yukawa to a Lectureship in Physics. No doubt at the instigation of Nagaoka, who knew both men, a meeting was arranged between Nishina and Yukawa. The crucial point came to Yukawa in October 1934. The nuclear force is effective at extremely short distances. My new insight was that this distance and the mass of the new particle are inversely related to each other.

In the early 1930's it was generally considered that protons and neutrons made up the atomic nuclei. These were surrounded by rotating electrons, and by neutrinos. Heisenberg had argued that the force inside the nucleus should be given by a force between pairs of nucleons, in which the potential increased as $1/r$ over an infinite range. But Yukawa's suggested potential was of the form $(exp. (-kr)/r)$, operating over the short range ($1/k=h/ur$) where "h" is Planck's constant and u is the meson mass. By using the nuclear force range roughly known to be 2 10^(-13) cm. the mass came out to be about 200 times the electron mass.

In 1937 experimental confirmation of this was obtained with cosmic rays.

Erwin Rudolf Josef Alexander Schrodinger (12/8/1887—4/1/1961)

Erwin Schrodinger was an only child in an indulgent family, and may have believed that the world revolved round him. His mother Georgina Bauer married Rudolph Schrodinger in 1886. Rudolph had inherited a small linoleum and oilcloth factory. They were married in the Lutheran Stadtkirche, the bridegroom being a Catholic, and the bride a convert to the Evangelical Church. Thus Erwin was nominally a Protestant, despite living in mainly Catholic Austria.

Erwin was born in Vienna in 1887, and was baptised by the Minister of the Evangelical Church. He received his mother's full attention, as well as that of the young maids and nurses, all of whom considered Erwin to be a budding genius. From his aunt Minie he learned to speak English before he spoke German well. He kept a diary of daily happenings all his life. Although his father had trained as a chemist, he had a great interest in Italian painters, and was able to do landscape drawings and etchings. And to his son he was a friend, teacher and inexhaustible conversation partner.

Entrance to the Gymnasium was at age 9 or 10, and Erwin had lessons at home two mornings a week, instead of going to elementary school. He entered the Gymnasium at age 11, after tests in Greek and Latin.

The family never entered a Church, except to be married or buried. But Erwin had a great respect for saints and mystics. He learned his negative attitude to religion from his father, though his mother was mildly religious. Erwin himself was an enthusiastic Darwinist. He was also greatly influenced by Schopenhauer (1788-1860) who kept a copy of the Hindu Scriptures at his bedside. Schopenhauer also kept a statue of the Buddha dressed as a beggar. His philosophy is closely related to the wisdom of the east, and many people learned of Vedanta and the Upanishads from his writings. He influenced Schrodinger greatly. Schopenhauer often called himself an atheist as did Schrodinger, and if Buddhism and Vedanta can be described as atheistic religions then they are right. They both rejected the idea of a personal God.

In 1946 Schrodinger attended one of Carl Jung's Conferences on "The Spirit and Nature". Schrodinger himself spoke on "The Spirit of Science", in which he said, "Spirit evades objective examination, because it is not an object but a subject." He quoted a Vedanta aphorism which says, "Subject and Object—the I and the Not I—are in their essence opposed to each other, like light and darkness." The great Indian philosophers were concerned only with the Ego, that consists of Thought and its relationship to the Godhead. Schrodinger wished to identify this Ego with Spirit. Science can examine only the object—the non-self.

In the Bhagavad-Gita teachings of three paths to salvation: (1) the path of devotion, (2) the path of works and (3) the path of knowledge, Schrodinger chose the path of knowledge.

In 1917 Rudolph Schrodinger's linoleum and oilcloth factory closed. Erwin's military pay stopped at the end of 1918, and the family in Vienna often ate at a community soup kitchen over the winter. But by 1919 he was able to marry his wife Amy, the daughter of a Court Photographer at Salzburg. They were married twice, first in the rectory of the Catholic Church in the parish where Amy stayed, and secondly in the Evangelical Church. The marriage lasted until death. Amy was 23 and Erwin 32. Amy was kind-hearted. They had no children. In later life Amy helped him to find other feminine companionship.

In 1920 Schrodinger accepted an Assistantship to Max Wien in Jena, but then moved to Stuttgart as Associate Professor, before becoming a full Professor in Zurich in1921, where his teaching load was 11 hours per week in 1922/23. It was in Zurich that he found in Hermann Weyl a kindred spirit, who was sympathetic with Schrodinger's interests in Eastern religions, and especially with the concept of Tantrism, in which all created activities in the cosmos are allied to the creativity of human erotic experiences. In 1923/24 Zurich citizens organised soup kitchens in Stuttgart, to feed thousands every day, because the Allied blockade of Germany had been maintained after the Armistice and thousands perished in German cities.

The Discovery of Wave Mechanics

Schrodinger's Head of Department in Zurich was Professor Debye. He had read a paper of De Broglie's in which the subject of matter waves was dominant, but did not understand it. So he asked Schrodinger to give a talk on the subject. In his first paper on the subject Schrodinger acknowledges help from Hermann Weyl for indicating the way to solve the equation. The equation itself was derived by Schrodinger, and is therefore called Schrodinger's equation. The question is not infrequently asked, "Where did Schrodinger's equation come from?" And the answer given by Richard Feynman is, "It came out of Schrodinger's head." When Schrodinger wanted to work without any distractions he used to place a pearl in each ear to shut out the noise, to which he was very sensitive. But at this time Schrodinger was also going through marital difficulties. His marriage with Amy was at a high point of tension with constant talk of divorce. And according to Hermann Weyl, Schrodinger's best work at this time was done during a late erotic outburst.

Schrodinger's first of six papers on Wave Mechanics has been described as one of the greatest achievements of twentieth century physics. By 1960 more than 100,000 papers had appeared based on applications of the Schrodinger Equation. Its solution was given in terms of the amplitude "u" of the wave function of the electron. He imagined this "u" function to be vibrating at high frequencies "hf" given by the differences of the energies associated with the Bohr conditions. His fame spread worldwide and in the following year he was invited to America where he gave over fifty lectures in three months. After his return to Zurich he was offered and accepted the offer of the Chair in Berlin, previously held by Planck, and he took this up in October 1927.

Schrodinger visualised the motion of atomic electrons as being governed by a system of generalised three dimensional waves, surrounding the atomic nucleus, whose shape and vibration frequencies were determined by the field of electric and magnetic forces. The solution to his equation allowed him to calculate how his quantum probability waves move, which could be compared with experiment. His lectures were good, but he did not encourage any who wanted to become his research students. His advice to all such was, "In first year do nothing but mathematics, in second year do the same, and in the third year come and discuss a research problem." But even here apart from his fellow student Frenzel, he never in all his life had a close personal male friend. He never had the experience of brothers or sisters whom he could call his own, nor even of students for whom he could be an intellectual father and mentor. But the Schrodingers moved to Berlin when the German economy was prospering, and partying was common. While not among these experiences Erwin fell in love with the young wife of a physics colleague in Innsbruck. He had one other mistress as well

Early in 1929 the German economy began to falter. The Wall Street crash took place in October of the same year, and there were street battles in Berlin between Communists and Nazis. There were riots in the University and some classes were closed. The students agitated especially for a quota limitation on enrolment of Jews. Erwin made no secret of his dislike of the Nazis. In the national elections of July 31 the Nazis became the largest party in the Reichstag. But most German scientists considered the public expression of any public allegiance to be inconsistent with the dignity of their profession. The hierarchy of Professors was not only anti-Semitic, it was also anti-socialist, and in Prussia anti–Catholic.

It is only fair to mention that this anti-Semitism was not confined to Germany. In the 1930's a University like Harvard restricted its intake of Jewish staff and Jewish students. But in Germany March 31 was declared to be a day of National Boycott of the Jews. Erwin was appalled by the mob action, remonstrated with a stormtrooper and was fortunate to be saved by a younger colleague. The Prussian Academy was forced to expunge the names of Einstein and Schrodinger from its membership records.

The Professor of Physics at Oxford University, F. A. Lindemann, had been appointed in 1919. His mother was an American and his father an Alsatian. He had been educated under Planck in Berlin, showing early promise in research. Lindemann took the view that he would like to have the talents of one or two outstanding theoretical physicists for Oxford University. But he soon realised that more than one or two would require help, and so he discussed the problem with the Academic Assistance Council, chaired by Lord Rutherford. This Council had limited funds which had been raised by public subscription, about £13,000. Also because of the economic recession many young British scientists could not find employment. Lindemann approached the chairman of I.C.I. who promised additional funds specifically for refugees, so that no British scientist would be disadvantaged. Only scientists of established reputation would be offered places. Then Lindemann went to Berlin to draw up a list of those who might be forced to leave Germany. He had tea at Schrodinger's house, and discussed the possibility of a place for Fritz London who was a research Fellow at the University, and Schrodinger's assistant. London had asked for time to think it over, at which

Schrodinger said he would go if London would not. In fact both went. Lindemann himself was not Jewish, but he was a man of great courage and determination. As an example of his courage he was the first person to demonstrate how to get an aircraft out of a spin. He himself had worked out mathematically that it required the aircraft heading towards the ground to have its speed increased as it sped downwards. Then he took an aircraft up, put it into a spin deliberately, and as it spun towards the ground increased its speed sufficiently to get it out of the spin into which he had put it. Many lives were saved through this mathematical discovery.

As a result of Lindemann's action Schrodinger was given a temporary Fellowship at Magdalen College, Oxford from 1933-36. During this period he was considered for a Chair at Edinburgh University, but either due to administrative delays or his unwillingness to have to lecture to first year students, this post went to Max Born, who as a Jew was dismissed from his Chair at Gottingen. Born, later in 1954, was awarded the Nobel Prize himself. Meanwhile President de Valera in Ireland had it in mind to set up an Institute of Advanced Studies in Dublin. He himself was a respected mathematician. And so Schrodinger travelled to Dublin in 1939, after a temporary stay in Austria at the University of Gent.

It was in Dublin that he wrote his influential book *What is Life?* It has been described as one of the most influential writings in this century. It attempts to comprehend some of the genuine mysteries of life, and has been responsible for J. B. S. Haldane, Francis Crick and his colleague Watson choosing to enter the field of biology. And it was at King's College, London, that the first double helix photograph was obtained, under the headship of Professor John Randall. And as an example of breadth of interests one only has to recall that it was he who designed the microwave cavity magnetron, one of the great inventions of the Second World War.

As examples of Quotations by Schrodinger the following are noted:

(1) Thus, the task is not so much to see what no one has yet seen, but to think what nobody has yet thought, about that which everybody sees.
(2) I don't like quantum mechanics, and I'm sorry I ever had anything to do with it.

Hermann Weyl (1885-1953)

Hermann Weyl was born near Hamburg where his father was Director of a bank. At age nineteen he entered the University of Munich to study Mathematics and Physics. He then transferred to the University of Gottingen, possibly because of the reputation of Professor Hilbert there. Mathematics had been established at Gottingen by Carl Friedrich Gauss, who had been Professor there for about fifty years until his death in 1855. Later Felix Klein, an applied mathematician, and David Hilbert, a very pure mathematician, were appointed Professors. Klein succeeded in having Prussia establish a series of research institutions in applied sciences, of which several were in Gottingen. He visited major industrial establishments once a year, and this was usually followed by a banquet. Unfortunately, one day Klein fell ill after everything had been arranged, but since the industrialists included some very important people, another celebrity had to be found as the banquet speaker. Professor Hilbert, the pure mathematician, agreed to prepare a speech, and was carefully coached to say kind words on the relationship between mathematics and technology. When the time came he said, "Gentlemen, mathematics and technology are on the best possible terms with each other. They are so now, and will remain on the friendliest terms in the future. For the simple reason, gentlemen, they have nothing to do with each other." It is clear that Professor Hilbert's views had neither influenced past decisions in Gottingen nor did they influence all of his students. With regard to the immediate past it is recorded that in 1905 the University of Gottingen conferred on Oliver Heaviside the degree of Ph.D. *honoris causa.* The felicitous phrasing on the diploma, translated from the Latin, has already been given as:

That Eminent Man
Oliver Heaviside
An Englishman by Nation, Dwelling at
Newton Abbot
Learned in the Artifices of Analysis
Investigator of the Corpuscles
Which are Wont to be called Electrons
Persevering, Fertile, Happy though

Given to a Solitary Life
Nevertheless among the Propagators of the
Maxwellian Science
Easily the First

No British University ever gave Heaviside an honorary degree.

With regard to the effect on Hilbert's students it is pleasing to record that one at least, in the person of Hermann Weyl, although predominantly also a pure mathematician, did become the first person to solve the problem of radio propagation over a lossy flat earth in 1919. But prior to this in 1908 he obtained his doctorate for work on singular integral equations, and in 1913 published his first book on Riemann Surfaces, a book that was republished in 1997 from the 1913 edition.

In the same year he married a girl from a Jewish background, and was also appointed to a Chair at the Zurich Technische Hochschule. There he came into contact with Einstein who was developing his General Theory of Relativity at the time. Weyl remained at Zurich until 1930 when he was appointed to be Hilbert's successor in Gottingen. It was probable that he would have remained at Gottingen permanently had it not been for the German hostility to the Jews. But in 1933 he accepted a post at the Princeton Institute for Advanced Studies, where Einstein had also gone for the same reason. Weyl retired from Princeton in 1952, spending part of his time there and part in Zurich, where he died in 1955.

In his professional life Hermann Weyl wrote over one hundred and fifty research papers and over a dozen books. Most of these were written in German, but a small number have been translated into English. In a self-deprecatory way Weyl himself wrote, "The gods have imposed upon my writing the yoke of a foreign language that was not sung at my cradle."

Paul Adrien Maurice Dirac (8/8/1902—20/10/1984)

Paul Dirac was the son of Charles Adrien Dirac, a Swiss citizen who had been educated at the University of Geneva, and had then come to England to teach French in Bristol. His mother, Florence, came from Cornwall and worked in the University Library. There were three children of the marriage, the eldest boy Reginald Dirac, then Paul, and lastly the young sister Beatrice

Dirac. Paul showed mathematical ability even in Primary School. At age twelve he entered Secondary School, and because of the war found that a number of senior boys, having joined the forces, left more freedom for younger boys to use the science laboratories.

At age sixteen he studied electrical engineering at Bristol University, obtaining his degree in 1921. Thereafter he had industrial experience for some months. But probably because of his inability to interact verbally with his colleagues this came to an early end. He was then fortunate to be given the chance to study Mathematics at Bristol University without the payment of fees. In two years he was awarded first class honours. This enabled him to enter Cambridge University in 1923, in the hope that he could research under Ebenezer Cunningham, an expert in relativity. This turned out not to be possible, and instead he was taken on by Ralph Fowler who was an expert in statistical mechanics, and was also knowledgeable in the quantum theory of atoms. Within six weeks of starting research Paul Dirac had written two papers in statistical mechanics, and within another four months his first paper on quantum problems, followed by another four papers by mid-1925. Yet within this period of most productive research, Dirac suffered the tragedy of his brother's suicide. It appears that the result of this was a separation from his father, since when he was awarded the Nobel Prize in 1933 he did not invite his father to the ceremony.

To answer the question of what was the merit in Dirac's work which has made him so famous we quote from an observation by Feynman: "Dirac got his answers by guessing an equation." This is analogous to Feynman's answer to the question, "Where did Schrodinger's Equation come from?" To this his answer was, "Schrodinger's Equation came out of Schrodinger's head." For the relativistic version of quantum mechanics the solution must satisfy the relativistic relation between the energy E and the momentum p of a particle. This says that the square of the energy E is equal to the square of (the momentum times the velocity of light) plus the square of the (mass times the velocity squared of light). When the square root of both sides is taken, both positive and negative square roots are obtained. Then as Hey and Walters explain, "Dirac suggested that negative energy levels did exist, but were already occupied by electrons." Then, because of Pauli's Exclusion Principle, no ordinary positive energy electron can make a transition to any of these levels. Thus a quantum box, which is apparently empty, containing no positive energy electrons, in fact has a fully occupied "sea" of negative energy electrons.

Now we know that if we shine light on an atom, electrons can absorb energy from a light photon and jump to an excited state. What happens if we shine light on an empty box? According to Dirac we should be able to excite one of these negative energy electrons up to a positive energy state. Instead of an empty box we would then have a positive energy electron, together with a hole in the Dirac Sea. Relative to the normal empty box state, a box with a hole in the sea is lacking some negative energy and negative charge. Compared with the original state, a hole in the sea has positive energy and positive charge. The physical process we have described is the creation of an electron-positron pair by a photon. The positron is the "antiparticle" of the electron, a particle with the same mass but opposite charge.

The positron was found experimentally in 1932, and the antiproton, the antiparticle of the proton was discovered in 1955. Its discovery came later because it had to wait for an accelerator which could provide enough energy for the creation of proton-antiproton pairs. And the converse to pair creation was also predicted by Dirac. If we have a positive energy electron plus a positron hole in our quantum box, then the electron can jump back into the sea and fill up the hole, leaving an empty box together with two photons to take away the annihilation energy.

The later feature of relativistic quantum mechanics, and one which is illustrated by both the pair creation and annihilation processes, is the possibility of transforming energy into matter, i.e. the number of quantum particles can change. Since Dirac relies on the Pauli principle to prevent positive energy particles jumping to a lower energy state, this filled sea trick will not work for bosons. There is a boson called a pi-plus, which has an associated pi-minus as its antiparticle. The Dirac sea with its apparently infinite negative charge and mass, disappears from a proper many-body quantum theory, which allows for particle creation and annihilation processes right from the outset. Such many-body theories are examples of "Quantum field theories". The relativistic quantum field theory describing the interactions of electrons and photons is known as quantum electrodynamics or QED for short. QED combines Maxwell's Equations of electromagnetism, quantum mechanics and relativity. It is the most successful theory physicists have yet constructed and it has been tested to an astonishing accuracy. To demonstrate that this is no idle boast, consider the spin of the electron that caused Schrodinger so much trouble. The spinning electron acts like a little magnet and the strength of the magnetic moment of the electron can be calculated in QED. The result may be expressed in terms of the "g factor" of the electron. Its classical value is two, but the value predicted by QED is 2.002319 which is in complete agreement with the experimentally measured value. To go further into this involves advanced mathematics. But Feynman's pictorial approach avoids this in an intuitive way.

Before introducing this approach, however, the section on Dirac will be concluded with some of his quotations:

(1) If one is working from the point of view of getting beauty into one's equations, one is on a sure line of progress.
(2) In science one tries to tell people, in such a way as to be understood by everyone, something that no one ever knew before. But in poetry it is the exact opposite.
(3) God used beautiful mathematics in creating the world.
(4) Mathematics is the tool specially suited for dealing with abstract concepts of any kind, and there is no limit to its power in this field.

Against the implications of (3) above it has to be said of Dirac, as quoted by Pauli, "Our friend Dirac, too, has a religion, and its guiding principle is There is no God and Dirac is his prophet." But this is unfair to Dirac. His concern for a fair assessment of a man's reputation was shown when he heard a fellow physicist pour scorn on Eddington's book *Fundamental Theory*. Dirac's reply was, "One must not judge a man's worth from his poorer work; one must always judge him by the best he has done."

There is also a human account of Dirac, as recorded in a newspaper interview he gave to a Wisconsin journalist. The local journalist recorded it as follows:

'I been hearing about a fellow they have up at the Uni this spring—a mathematical physicist, or something they call him— who is pushing Sir Isaac Newton, Einstein and all the others off the front page.—His name is Dirac and he is an Englishman.— So the other afternoon I knocks at the door of Dr Dirac's office in Sterling Hall and a pleasant voice says 'Come in.' And I want to say here and now that this sentence 'Come in' was about the longest one emitted by the doctor during our interview.

I found the doctor a tall youngish looking man, and the minute I see the twinkle in his eye knew I was going to like him—he did not seem at all busy. Why, if I went to interview an American scientist of his class—he would blow in carrying a big briefcase, and while he talked he would be pulling lecture notes, proofs, reprints, books, manuscripts or what have you out of his bag. Dirac is different. He seems to have all the time there is in the world, and the heaviest work is looking out the window.

"Professor," says I, "I notice you have quite a few letters in front of your last name. Do they stand for anything in particular?"

"No," says he.

"Fine," says I. "Now Doctor, will you give me in a few words, the lowdown on all your investigations?"

"No," says he.

I went on. "Do you go to the movies?"

"Yes," says he.

"When?" says I.

"In 1920—perhaps also 1930," says he.

"And now I want to ask you something more. They tell me that you and Einstein are the only ones who can really understand each other.—Do you ever come across a fellow, that even you cannot understand?"

"Yes," says he.

"Do you mind releasing to me who he is?"

"Weyl," says he.

Now Weyl was a very pure mathematician, and Dirac was a theoretical physicist.'

Richard Phillips Feynman (11/5/1918—15/2/ 1988)

Richard Feynman was the son of both a Jewish father and a Jewish mother. His father was born in Minsk, in White Russia, and his mother in the United

States, though she was of Polish descent. They were married in 1917 and their first son Richard was born in Manhattan in the following year. Richard's father was a man who loved science, and it was his desire that his son should follow this profession, which he had not had the opportunity of doing himself. Under this influence it is recorded that Richard learned a great deal of science from the *Encyclopaedia Britannica*, taught himself elementary mathematics before he met it at school, and experimented with electricity at home. In Secondary School in his final year he won the New York University Maths Championship. But he found no interest in non-scientific subjects.

Because of a quota system for Jews which operated in the United States at the time, entry to University was not straightforward. But he was admitted to the Massachusetts Institute of Technology in 1935 at age seventeen. This was to study Mathematics originally, then there was a brief spell in Electrical Engineering, and finally he transferred to Physics. Fortunately this was early on, and in his second year he was able to take "Introduction to Theoretical Physics" which was a course intended for graduate students. Moreover because there was no course provided in "Quantum Mechanics" he and a fellow student self-studied this subject both during term time and in the summer vacation of 1936. In 1937 Feynman was studying Dirac's *The Principles of Quantum Mechanics*.

At the end of his four years study at MIT Feynman decided to apply for research. On the recommendation of his Head of Department he went to Princeton University, where his supervisor was John Wheeler. Princeton was, of course, where the Institute of Advanced Studies was located, and among those attending Feynman's first Seminar were Einstein, and Pauli.

Feynman went on to develop a new approach to quantum mechanics, and it was soon realised that Wheeler's ability in mathematics was not up to the standard of Feynman's. After two years' work Feynman transferred to the Manhattan Project at Princeton, where he developed a theory for separating Uranium 235 from Uranium 238. Meanwhile Wheeler went to Chicago to work on the first nuclear reactor with Fermi. In 1942 Feynman received his doctorate from Princeton, and shortly after this he married his girlfriend of many years. She had recently contracted tuberculosis, and was to die within three years. But he demonstrated his commitment personally as he always did technically.

After 1945 Feynman took up a teaching job as professor of Theoretical Physics at Cornell University. The stress of his bereavement meant that five years were to elapse before he could take up research again. When he did so he went back to the quantum theory of electrodynamics that he had been working on for his doctoral work. We will now examine the development of this under its abbreviated title QED, from the combined contributions of Sin-Itiro Tomonaga, Julian Schwinger and Richard Feynman, who shared in the award of the Nobel Prize in 1965.

Quantun Electrodynamics (Q.E.D.)

The work of these three Nobel Prize-winners was carried out in about 1948, quite independently of each other. In two cases, those of Schwinger and Feynman, it was carried out in America, but in the third case of Tomonaga, in Japan. After the war with Japan ended in 1945, serious food shortages developed there. In the case of Tomonaga, this had the effect of his being unable to be engaged in anything that required thinking. So he decided to translate his Japanese papers into English.

In April 1946 Tomanaga gathered all his young research people together. He wished to involve them in a combined attack on quantum electrodynamics, following his own earlier researches in that area. As part of the process of welding them into a harmonious group they would meet once a week to discuss progress.

Then in 1947 Tomonaga obtained his first information on the Lamb Shift experiment, carried out at Columbia University by Willis Lamb and R.C. Retherford. This revealed a small difference in energy between two energy levels of the hydrogen atom in quantum mechanics. According to the Schrodinger equation these two energy levels should have been the same. But the Lamb shift became one of the foundations of quantum electrodynamics. And along with this experimental evidence there came a theoretical paper of Bethe's in which he made a nonrelativistic calculation of what this frequency shift would be. He found it to be 1040 MHz, close to the experimental result.

In his speech introducing the Nobel Laureates, Ivor Waller put it thus: "As soon as Tomonaga knew about the Lamb Shift experiment and Bethe's paper he realised that an essential step to be taken was to substitute the experimental mass for the fictive mechanical mass which appeared in the equations of quantum electrodynamics, and to perform a similar renormalisation of the charge. The compensating terms which had then to be introduced in the equations should cancel the infinities."

Earlier History of Japanese Electrical Research

Tomonoga (1906—1979) was born in Tokyo, the son of a philosophy professor there. His family moved to Kyoto in 1913, and he and Yukawa became close friends both at school and at University there, which they entered in 1923. On graduation in 1929 at the start of the economic depression, Tomonaga decided to stay in Kyoto as an unpaid assistant. This lasted for three years.

In 1931 both Tomonaga and Yukawa attended a lecture given in Kyoto by Yoshio Nishina (1890-1951). Now Nishina was the founder in Japan of experimental and cosmic ray research. He also supervised the construction of accelerators in Tokyo. At the end of the lecture both Tomonaga and Yukawa asked more questions than anybody else, and Nishina offered a paid post to Tomonaga in his laboratory in Tokyo. Soon Tomonaga became the chief theoretician. It was run very informally, and described as follows: "The Nishina Laboratory was full of freshness. We all got together after lunch every day discussing maths, physics and even beer parties."

During the summer of 1935 the theoreticians got together and translated Dirac's book. This was found to be heavy going, and the load was found to be eased by not working on Sundays.

In 1937 Tomonaga went to Leipzig to work with Heisenberg, and on the basis of the two years' work he did there he was awarded a D.Sc. by Tokyo on his return. Theoretical work he did in Leipzig involved the discrepancy between the measured lifetime of the mesotron and the calculated values. During this time in Leipzig Yukawa kept him posted on mesotron theory, and this interested Heisenberg. But after two years in Leipzig war broke out and on September 1st 1939 Tomonaga boarded a ship in Hamburg for his journey home. In 1940 he was appointed professor of Physics at Tokyo College of Science and Literature, where his gifts as a teacher became legendary. The story is told that on one occasion he gave a seminar which lasted several hours, during which he wrote down several long equations without ever referring to any notes.

After 1948 Tomonaga took the view that since everybody was now doing quantum electrodynamics research it was time for him to leave this fashionable field. He died in 1979.

Julian Schwinger (1918-1994)

Julian Schwinger's father came from a middle class Jewish family. He had been born in a village in the foothills of the Carpathian Mountains and came to the United States around 1880. His skills lay in the design of women's clothing. Julian's mother came from Lodz in eastern Poland. She was also Jewish and her parents were associated with clothing manufacture.

Their son Julian was exceedingly precocious and attended Townsend Harris which was New York's premier High School for gifted children. He entered that school in 1932 and graduated from there to enter the City College of New York in 1934. But he did not do well there because he spent too much time reading advanced physics and mathematics. He was described as "very, very, shy, introverted, gentle, kind and musical". During the Fall of 1934 Julian worked on the quantum mechanical description of the behaviour of spin in magnetic fields. And before he graduated as a Bachelor of Science in 1936 he had already written his Ph.D dissertation.

His Head of Department, I. I. Rabi (1898-1989) then got a travelling scholarship for Julian which would allow him to visit Wisconsin for six months and then proceed to California for another six months. To fulfil his duty as Head of Department and also towards Julian, Rabi visited Wisconsin after a suitable interval, to find that Julian was satisfied, but the Department at Wisconsin had hardly seen him there. Further enquiries elicited the information that Julian was working during the night and sleeping during the day. It appeared that Julian believed that the staff at Wisconsin wanted him to work on a particular project, and he had his own idea of what he should work on. This practice of being a night worker remained with Julian throughout his life, and of course it upset the workings of departments. But Julian was a special case and his idiosyncrasy was accommodated. Another example of this problem relates to Julian taking a course in Statistical Mechanics in 1938/39. The lectures were given by a visiting Professor from Holland, whose method of assessment involved an individual oral examination. Julian surprised the examiner by turning up for it at ten o'clock at night, but was required to return at ten o'clock on the following morning.

In February 1942 Bethe proposed the formation of a team of five theoreticians, all Ph.D's which would be headed by Schwinger. Accordingly Schwinger arrived at M.I.T. in late 1943, after declining an offer earlier in the year to work at Los Alimos on the atomic bomb.

Schwinger was a perfect lecturer. When he was asked to lecture at an American Physics Society in 1948 the lecture had to be given in triplicate because it was so good. Also when I. I. Rabi was unavoidably away at Columbia University he would ask Julian to take his class on Quantum Mechanics, knowing it would be better done than when he gave it himself.

Shelter Island and its Aftermath (June 1947)

A former president of the New York Academy of Sciences, Duncan MacInnes, proposed a small conference of top people on "Fundamental Problems of Quantum Mechanics". He was aware that conferences were becoming too large, and so he wanted to keep the numbers down to between twenty and thirty. The invited participants included Oppenheimer, Teller, Rabi, Wheeler, Schwinger and Feynman.

Physics was popular, but more important to the participants was that the Dirac theory appeared to be breaking down. The reports of the Columbia University experiments by Lamb and Retherford on the first day of the conference made a deep impression on Schwinger. Previously the theoreticians were trying to get rid of infinities in their calculations, but now here was a finite frequency shift of about 1000MHz, which could be and was measured.

The following year saw a repeat of a small group of top scientists meeting together at Pocono. Schwinger and Feynman contributed in their own characteristic ways. Of Schwinger's style of lecturing it was said, "Other people publish to show how to do it, but Schwinger publishes to show you that only he can do it."

Feynman's contribution was based on his observation that the probability of an electron or a photon proceeding from a source S to a detector D, in the absence of any screen between them, was the sum of amplitudes for all possible paths between S and D. His application of this to practical problems involves the use of diagrams, which like the mathematics deliberately avoided in this book will also be omitted.

But possibly the best introduction to Feynman's thinking is given in his eleven-page Nobel Prize acceptance speech in Sweden in 1965. In that lecture Feynman goes over the history of his own ideas from 1939-1947. This is real history covering the blind alleys in which time is spent, before another forward step is taken. He had read the last sentence in Dirac's book on *Quantum Mechanics*, which said, "It seems that some essentially new physical ideas are here needed." Two infinities were the cause of the difficulties at that time. The first was the infinity energy of the electron with itself. The second was the infinite number of modes associated with a given volume of field as the frequency increased. But Feynman argued that there was no field, but simply a delay between the motion of one charge and its effect on another. "The sun atom shakes; my eye electron shakes eight minutes later."

It appears that Feynman's type of analysis is likely to be used more in the future than that of Schwinger.

Lightning Source UK Ltd.
Milton Keynes UK
02 March 2011

168507UK00002B/25/P